Work in a Warming World

Carla Lipsig-Mummé
Stephen McBride
Editors

Queen's Policy Studies Series
School of Policy Studies, Queen's University
McGill-Queen's University Press
Montreal & Kingston • London • Ithaca

Publications Unit
Robert Sutherland Hall
138 Union Street
Kingston, ON, Canada
K7L 3N6
www.queensu.ca/sps/

The preferred citation for this book is:
McBride, S., and C. Lipsig-Mummé, eds. 2015. *Work in a Warming World*. Montreal and Kingston: Queen's Policy Studies Series, McGill-Queen's University Press.

Library and Archives Canada Cataloguing in Publication

Work in a warming world / edited by Carla Lipsig-Mummé and Stephen McBride.

(Queen's policy studies series)
Includes bibliographical references and index.
Issued in print and electronic formats.
ISBN 978-1-55339-432-7 (pbk.). – ISBN 978-1-55339-433-4 (epub). –
ISBN 978-1-55339-434-1 (pdf)

1. Sustainable development. 2. Sustainable development – Canada. 3. Sustainable development – Government policy. 4. Climatic changes – Government policy. 5. Climatic changes – Economic aspects. 6. Job creation. 7. Green movement. I. Lipsig-Mummé, Carla, author, editor II. McBride, Stephen, author, editor III. Series: Queen's policy studies series

HC79.E5W58 2015 338.9'27 C2014-907703-3
C2014-907704-1

For John and for Claire
Carla Lipsig-Mummé

For Jan, Emily, Morna, and Shona
Stephen McBride

TABLE OF CONTENTS

LIST OF TABLES, FIGURES, AND APPENDICES

TABLES

FIGURES

APPENDICES

I

TRENDS AND CHALLENGES

INTRODUCTION

STEPHEN MCBRIDE AND CARLA LIPSIG-MUMMÉ

Among global warming's many effects is a major impact on employment and work. The extent of global warming will create some employment opportunities and eliminate others. The same is true of policies put in place to deal with climate change. Such policies are often articulated in terms of their environmental consequences, but the economics of the policy responses receive scant attention.

In part, this is because the interests of the economic and the environmental are often counterposed, with environmentally friendly measures receiving greater public support in economically prosperous periods, but regarded as dispensable in times of economic hardship. Continued public support for climate-change-control measures may, therefore, depend on them being embedded in measures that ensure employment and, with it, economic security. As well, work activities in developed countries are responsible for most greenhouse gas (GHG) emissions. Since GHG emissions generated by human activity are implicated in most global warming, what happens in the world of work, employment levels aside, has an impact on how quickly the world is warming. Thus, if measures to maintain employment are necessary to politically sustain the climate-change-control model that emerges, so too are measures to enable a greener labour process so that a given level of work activity produces over time a progressively smaller amount of GHG emissions.

One common theme in this volume is that environmental and economic-employment policies exist in separate silos, each proceeding without much reference to the other. This theme is the product of a five-year collaborative research project called Work in a Warming

Work in a Warming World, edited by Carla Lipsig-Mummé and Stephen McBride. Kingston: School of Policy Studies, Queen's University.

World: Adapting Canadian Employment and Work to the Challenges of Climate Change, based at York University in Toronto, whose purpose is to better understand the present and potential role of work in Canada's transition to a low-carbon economy.[1] The contributors share the normative perspective that such mutual blindness needs to be overcome. The environmental issues involved in mitigating and adapting to climate change cannot be overcome unless they rest on an economically secure basis. And economic problems, long-term as well as crisis-induced – as in the aftermath of the 2007–08 financial crisis – cannot be dealt with in a way that sacrifices attention to environmental challenges, particularly those posed by climate change. The two need to be bound together in a symbiotic and mutually supportive way.

That said, getting to such a virtuous and non-zero-sum position will not be easy. How can climate policy move beyond its current employment blindness? The Intergovernmental Panel on Climate Change defines two primary policy strategies to slow global warming: mitigation – reducing the production of GHG emissions – and adaptation. For more than two decades, there has been competition to determine which policies have higher priority: those of mitigation or those of adaptation. The original focus of adaptation was defensive and reactive: develop policy and community-based measures to reduce vulnerability and strengthen community resilience in the face of inevitable global warming. More recently, the focus has moved from reactive adaptation to proactive adaptation: large-scale policy initiatives and measures that plan for and act ahead of climate risk and disaster. While the key actors in defensive adaptation were seen to be vulnerable communities aided by social policy, the state at national and sub-national levels is potentially the driver of proactive adaptation (Giddens 2011; Lipsig-Mummé 2013). The state's capacity to fund adaptation measures is compromised in those countries where austerity has been adopted as a response to the 2007–08 global financial crisis.

Mitigation has been the strategy favoured in global negotiations over the past two decades, as seen in international policy like the Kyoto Protocol, tactics like setting arbitrary goals for reducing carbon dioxide emissions, and the national and sub-national policy responses of most industrialized nations. Moves by governments in North America and the European Union (EU) have focused increasingly on market mechanisms (i.e., cap-and-trade schemes). Mitigation measures have the advantage of being cheaper and thus more affordable for fiscally constrained states.

But mitigation, to date, has not been effective enough. GHG emissions rose 5.9 percent in 2010, the largest increase since measurement began. Global economic governance has marginalized global environmental governance, and there is already evidence (see Chapter 1 by Sinclair and Trew) that this may further impede action by states and/or sub-national jurisdictions on climate, jobs, and work. Moreover, the failure to integrate employment and climate adaptation into mitigation impedes action on the climate change file, with potentially disastrous environmental consequences (Dupressoir et al. 2007).

A troubling pattern is emerging. The economic restructuring that results from grappling with climate change may merge with the effects of the financial and environmental crisis, which are predicted to have a long-lasting employment impact (Dupressoir et al. 2007). Policy measures taken in response to these challenges already have labour market consequences. Unless the employment impacts are an explicit part of the climate change decision-making process, they may get lost, with possible detrimental consequences not only for employment but also for the political support base for responses to environmental issues in a time of economic difficulty. The danger that environmental measures will be blind to employment, and vice versa, is already playing out in Ontario.

Indeed, as many of the contributors to the book make clear, the basic building blocks of an integrated economic and environmental strategy are absent and, therefore, need to be created – a process complicated by the need to incorporate the gender dimensions of the climate change issue into policy measures and create a vocational training regime that can contribute to green jobs and work practices. Further, existing institutional arrangements (whether inside Canadian federalism or around the world, as in the case of the architecture of global economic governance) may actively impede efforts to develop a rational and win-win approach to these problems.

The book is structured around two themes. The first theme surveys major trends and challenges for the world of work in engaging more effectively with the struggle to slow global warming. Five chapters discuss the squandered opportunities, silences, and neglected terrain in Canada's climate record. The second theme assesses efforts in professions, sectors, unions, and cities to integrate green priorities into the world of work and employment.

Scott Sinclair and Stewart Trew's chapter analyzes the international, national, and provincial impact of Canada's recent loss at the World Trade Organization (WTO). The dispute, brought by Japan and the EU against domestic content requirements in the *Ontario Green Energy and Green Economy Act* (informally called the Green Energy Act, or GEA), has refocused attention on the problematic role played by international trade treaties, including those signed under the auspices of the WTO, in limiting government flexibility in the area of environmental or sustainable development policy. Since the first WTO panel ruling on the case in December 2012, similar disputes against "buy local" policies on green power have been threatened against India, Europe, China, and the United States, putting into question the viability of domestic content requirements as an incentive for rapid development and deployment of renewable power. The chapter analyzes the WTO panel and Appellate Body decisions in relation to Ontario's GEA to assess the extent to which they constrain the ability of governments to combine job creation and environmental protection strategies. It also explores how emerging trade agreements – for example, between Canada and the EU – move the global goalposts further toward liberalization over localization in the area of renewable power. The chapter ends with recommendations for an increased role for the public sector in developing green energy to remedy the erosion of local industrial development.

John Calvert considers how labour in Europe has addressed the issue of global warming, with a specific focus on the role of the building trades in Germany and the United Kingdom. The built environment accounts for over 35 percent of energy use and GHG emissions in Europe. Reducing its carbon footprint is key to addressing global warming. The chapter reviews the EU's overall climate policy and examines national policies on greening the built environment. It then looks at some of the specific initiatives taken by the building trades unions to further the transition to a greener economy, including reviewing union policy prescriptions, specific union activities, and the direct role that building workers' unions have played in shaping the restructuring of their respective construction industries to deal with global warming.

The ability of unions to play a constructive environmental role depends on the broader policy framework established by governments and on their influence in their own industry. Where union density is high and unions are significant players in training and workforce development, as in Germany, they have been influential in shaping the industry's environmental policies. However, where unions are weak and have little say in training and workforce development, as in the United Kingdom, they have had little impact in greening construction. The role of labour is only one factor in the success of climate policies. But the presence – or absence – of union involvement does make a difference in the ability of countries to achieve a greener economy and society.

Marjorie Griffin Cohen highlights the gendered impact of climate change for both paid and unpaid labour as well as the way consumption can be differentiated along gender lines. Noting that what work has been done on these important issues has tended to be focused on the developing world, her contribution addresses the silence on gender in responding to climate change. It makes a start in understanding how gendered distinctions also operate in the developed world. She argues that knowing the gendered impact of GHG emissions can help develop appropriate policies for green jobs and a green economy.

Geoffrey Bickerton and Carla Lipsig-Mummé take up the neglected terrain of labour's role in climate strategy and ask, How can labour broaden and deepen its capacity to protect work and workers from the unique threats posed by climate change, all the while contributing to the struggle to slow global warming in the context of increasingly pessimistic climate science, a global economic crisis, a hostile national government, and strategic paralysis in the national and international political arena? The first and second sections of the chapter frame the issues. The third and fourth sections explore the challenges and dilemmas for labour leadership in the current political climate in Canada, presenting research undertaken by the Canadian Union of Postal Workers (CUPW) as a case study. CUPW's project on union initiatives in greening work in the postal sector, expanded to survey postal unions around the world, demonstrates the unexpected ways in which research carried out and shared through global union channels can catalyze international information sharing and action on greening work. The final section of the chapter asks, What would it take for Canadian unions to take up climate leadership in this critical moment?

In the final chapter in this first theme, Mark Winfield examines the role of renewable energy as industrial strategy – an issue of enormous importance to Canada. He takes Ontario's 2009 GEA as his case and sets it, and the difficulties it has encountered, within the international debate on the role of renewable energy as a catalyst for reversing deindustrialization in industrialized countries.

The act sought to achieve a number of policy goals. These included the rapid deployment of renewable energy resources in the province, in part to support the phase-out of coal-fired electricity generation. In addition, in the context of the ongoing decline of the province's manufacturing sector, accelerated by the impact of the 2007–08 financial crisis, the legislation was designed to encourage and facilitate the development of a competitive renewable–energy technology design, manufacturing, installation, operation, and servicing sector in the province. The government of Ontario consistently stated its goal of creating 50,000 new jobs in the renewable energy sector through the legislation and claims that 20,000 new jobs have in fact been created.

The actual employment and economic development impacts of the legislation are the subject of considerable debate. Winfield's chapter examines the key points of disagreement and contention over the economic impact of the Ontario legislation, particularly the Feed-in-Tariff (FIT) program established under it. It places these debates in the province in a wider comparative context, and it employs a discourse analysis framework to explore how similar discussions have unfolded in other industrialized jurisdictions pursuing large-scale renewable–energy development programs. The key case studies examine Germany, Denmark, Spain, the United Kingdom, and selected US states. Finally, the chapter draws out the implications of the analysis for the policy decisions that Ontario now faces with respect to its FIT program and wider electricity and energy programs.

The book's second theme is the greening of work by professions, sectors, unions, and cities. Kean Birch and Dalton Wudrich focus on infrastructure renewal and training engineers. They note the need to be forward thinking when it comes to core infrastructure because what is being built now will be with us for the next half-century or more. As a result, infrastructure planning, design, and development choices made today have a direct bearing on society's ability not only to mitigate but also to adapt to climate change in the future. It is therefore vital to look at whether and how climate change is being integrated into new and renewed infrastructure. What is evident is that it is increasingly being planned, designed, and built as sustainable infrastructure with an increased focus on resilience, adaptability, and social relevance. This represents a particular challenge for the engineering profession since engineers will be at the forefront of this effort, and they currently do not believe that they have the skills or knowledge necessary to react professionally to climate change. Thinking about the various implications for engineers of these changes (e.g., changing practices, changing standards, changing education and training) will enable engineers, policy-makers, and the public to better plan and adapt to climate change and avoid locking ourselves into inadequate or inappropriate infrastructure in the future.

Returning to the theme of training and preparation for climate change, John Calvert examines green training in the construction trades in Canada. This chapter focuses on the question of how effectively those working in the Canadian construction industry are addressing climate change. The transition to greener construction entails changes at every step of the building process, from initial planning to design and engineering to the actual work on the job site. The chapter focuses, more narrowly, on the role of the construction workforce in this transition. Successful implementation of green construction requires new skills, new training, new ways of working, and a new understanding of the potential of the built environment to address climate change. The question is whether – and to what extent – training and apprenticeship programs now include new green skills and whether on-the-job practices reflect best environmental practices.

While building workers are only one part of the larger construction process, international evidence suggests that their active involvement is vital to achieving Canada's climate objectives. However, their input has been marginalized in much of the industry. Consequently, a second component of this paper examines the role of unions, which still represent a significant part of the construction workforce. While this role varies, depending on occupation, construction sector, and region, unions have made a number of promising initiatives to incorporate green components into their apprenticeship programs. They have also lobbied for change in building codes, government procurement, and public policies.

Steven Tufts and Simon Milne contrast employer and union strategies for greening the hotel sector in Ireland, Australia, the United States, and Canada. They note that, in recent years, a number of labour union strategic initiatives have been developed that seek to leverage consumer preference against employers in the hospitality services sector. Some programs focus on rating and certifying hotels based on environmentally and socially responsible behaviour and labour-friendly practices. Bolder initiatives have attempted to link working conditions in the hospitality sector to the sustainability of the global food system. Campaigns are a response to the perceived "greenwashing" of hospitality through voluntary, self-reporting rating systems and inadequate food-sourcing policies. The chapter examines union campaigns that recommend hotels according to social and environmental criteria, and it attempts to link food security issues to food service production. Some of the campaigns discussed are Fair Hotels Ireland, the Green Star program (Australia), and UNITE HERE's Real Food Real Jobs campaign (United States and Canada). These emerging campaigns differ in orientation, but all face challenges in their ability to meet their strategic goals. There are questions related to the geographic scale of the campaigns and their ability to advocate for a more socially and environmentally sustainable hospitality services sector.

Stephen McBride, John Shields, and Stephanie Tombari look at the climate policies of cities, particularly mid-sized cities, and explore the discourse surrounding green economic and employment strategies. Canada's abdication from the Kyoto Protocol during the United Nations Climate Change Conference in 2011 sent a clear message from the federal government that the climate was not its concern. Despite the fact that Canada is one of the world's worst per capita GHG emitters, the Harper government has refused to adopt a national climate action strategy. And while Ontario, Quebec, and British Columbia have initiated a variety of progressive climate policies, the lack of national coordination has led to a Canadian climate response that is all over the map. Canadian municipalities are taking it upon themselves to respond to both climate and economic crises. The country's largest cities – Toronto, Vancouver, and Montreal – have been adopting policies to reduce GHG emissions produced by municipal services and buildings since the early 1990s. But the climate change policies of mid-sized Canadian cities have been largely ignored. This chapter explores conditions that lead mid-sized cities to pursue climate-friendly economic development policies, the ways cities learn about and adopt climate policies from other jurisdictions, and the various public-private partnerships that accompany green economics.

Megan MacCallum, Lindsay Napier, John Holmes, and Warren Edward Mabee complement this analysis by writing on renewable energy and sustainable jobs, focusing on the case of Kingston, Ontario, and the surrounding region. As part of a strategy to become Canada's most sustainable city, Kingston has supported renewable–energy generation projects in and around the municipality. An initial review of the region conducted in 2011 indicated that Ontario's GEA, combined with the presence of wind and solar resources, served to initiate a cluster of business activity. The initial report documented roughly 100 firms in the Kingston region that were engaged in developing renewable energy projects. Of these firms, however, most were very small, employing only a handful of employees, although each one indicated that it intended to grow over the next five years given the appropriate political and economic climate. In the summer of 2013, a follow-up survey was conducted to analyze progress toward these goals. It categorized green employment related to the renewable energy sector and tracked progress in each category of employment. It assessed the role of renewable energy as a driver for employment in the Kingston region as well as the potential for green energy created through these activities to support a broader category of green work.

CONCLUSION

In dealing with the challenges posed by climate change and economic crisis, we find considerable activity at the sub-national level – that is, the municipal, sectoral, and union-professional levels. To some degree, the message seems to have been heeded that being employment friendly is an essential route to environmentally friendly initiatives. These developments are encouraging. Clearly, however, such optimism is tempered by the broader context in which such activities occur.

As shown by the WTO case that struck down sections of Ontario's GEA, efforts to incorporate job creation into environmentally friendly policies are prone to the sanctions of international economic agreements created for an entirely different purpose. To the extent that climate change measures need to solidify political support by addressing the bread-and-butter issues of employment creation, these agreements actively impede that goal by prioritizing the various trade principles that constrain governments. Moreover, we can posit that sub-national, sectoral, and union-professional efforts would flourish under a supportive national policy regime but languish under one that is negative or indifferent. In the Canadian case, the latter situation obtains. It seems, too, from international comparisons that the strength of labour, and the degree to which its role is institutionalized in decision-making or consultative forums, has a significant impact on its ability to promote green working practices and employment. Where labour is weak, as in Canada, it can still play a role, but its overall influence in this, as on many other policy files, is limited.

One practical implication is that the institutional as well as ideational building blocks of an alternative that combines economic and environmental imperatives are still to be built. And time is short.

NOTE

1. Work in a Warming World is an Environmental Community-University Research Alliance of the Social Sciences and Humanities Research Council of Canada, funded from 2009 through 2015. More information is available at http://www.workinawarmingworld.yorku.ca/projects/work-in-a-warming-world/.

REFERENCES

Dupressoir, S. et al. 2007. *Climate Change and Employment: Impact on Employment in the European Union-25 of Climate Change and CO_2 Emission Reduction Measures by 2030*. Brussels: European Trade Union Confederation. https://docs.google.com/file/d/0B9RTV08-rjErMGstNWRPbkJWaWM/edit?pli=1.

Giddens, A. 2011. *The Politics of Climate Change.* 2nd ed. Cambridge: Polity Press.

Lipsig-Mummé, C. 2013. "Climate, Work and Labour: The International Context." In *Climate@Work,* ed. C. Lipsig-Mummé, 31–41. Black Rock, NS: Fernwood Publishing.

CHAPTER 1

INTERNATIONAL CONSTRAINTS ON GREEN STRATEGIES: ONTARIO'S WTO DEFEAT AND PUBLIC SECTOR REMEDIES

SCOTT SINCLAIR AND STUART TREW

INTRODUCTION

The *Ontario Green Energy and Green Economy Act* (informally called the Green Energy Act, or GEA) was a creative though by no means unique effort to address the climate change imperative, while at the same time revitalizing a faltering provincial economy. The 2009 act gave the production and use of renewable energy a notable boost, using a feed-in tariff (FIT) program borrowed from pioneering German and other successful European models to provide above-market rates for different forms of green power. To qualify for the FIT in Ontario, producers needed to source a minimum percentage of goods, services, or labour within the province. The premium rates encouraged new investment in renewables, while evidence suggests that the local content requirements (LCRs) helped create thousands of new jobs in the province.

These LCRs, more than the FIT itself, also attracted World Trade Organization (WTO) challenges from Japan and the European Union (EU), which both argued, successfully, that the Ontario policy illegally discriminated against foreign-produced wind and solar products in violation of the General Agreement on Tariffs and Trade (GATT) as well as the Agreement on Trade-Related Investment Measures (TRIMs). Canada insisted that the Ontario FIT with its LCRs was excluded from non-discrimination rules by virtue of being a type of government procurement, an argument the WTO dispute panel rejected. Canada appealed,

Work in a Warming World, edited by Carla Lipsig-Mummé and Stephen McBride. Kingston: School of Policy Studies, Queen's University.

but the WTO Appellate Body agreed with the panel, although for different and complicated reasons, which we will explain below.

The decision dealt an unfortunate blow to the feasibility, not to mention legality, of LCRs for renewable energy in a number of countries. For Ontario, it challenged the provincial government to salvage a landmark climate and green jobs strategy that was already under heavy attack by opposition parties, free-market think-tanks, and media commentators who believed that the GEA was too expensive and ineffective (McBride and Shields 2011). Ultimately and unfortunately, the provincial government appears to have acquiesced by removing domestic content requirements completely from future FIT and MicroFIT projects.

This chapter is divided into three parts. The first describes the basic features of the GEA and provides an early assessment of its successes. The second summarizes the WTO's dispute settlement panel and Appellate Body decisions in the Canada–Renewable Energy case (the "Canada Renewables" case) brought against Canada by Japan and the EU as well as its collateral damage to other FIT programs around the world that incorporate LCRs. Finally, we propose a path forward for Ontario that depends more on the public, as opposed to the private, sector to meet its climate and green economy objectives as it reforms the GEA to accommodate the WTO decision. In fact, we argue that a greater reliance on public power would not only make LCRs in renewable energy programs generally safer from trade challenges but also more manageable and cost effective.

Public procurement remains the most straightforward option for Ontario to achieve its complementary renewable energy and economic development goals. In the case of renewable energy contracts, where a public entity acquires the generation equipment, government would still be free under WTO rules to require that any portion of that equipment be manufactured in Ontario. There are promising signs that the Ontario government would like to move in this direction, at least partway. However, there is pressure in international trade agreements, notably the comprehensive free trade agreement between Canada and the EU, to bind sub-national governments, including municipalities and energy utilities, to procurement rules that prohibit LCRs on all purchases above certain low thresholds. In light of the WTO decision, and Ontario's efforts to use procurement to provide economic benefits to Ontarians, it will be essential for the province to fully safeguard its existing policy flexibility over renewable energy procurement under the EU deal and other, future trade agreements.

THE GREEN ENERGY ACT: A GOOD IDEA UNDER THREAT

Depending on whom you ask, the GEA of 2009 is either a victory for innovative sustainable development policy or an interventionist failure with high costs to taxpayers and, possibly, the environment (Winfield 2013).

The act has three main goals: to significantly increase the production and use of renewable energy in Ontario, to promote energy conservation, and to spur the creation of new green jobs in manufacturing and associated service sectors. Each of these aspects is integral to the overall success of the policy in reducing greenhouse gas (GHG) emissions, improving air quality, and stimulating local economic development.

The policy tool used to achieve these objectives is a FIT borrowed from the highly successful German model, which is estimated to have created in excess of 300,000 jobs while making Germany a green technology powerhouse (McBride and Shields 2011). In Ontario, producers of approved green energy from solar, wind, biomass, biogas, and landfill-gas projects are guaranteed a fixed (usually 20-year) above-market rate by the Ontario Power Authority (OPA), which purchases and redistributes the energy to the grid. The prices vary, depending on the technology, from 10.3 cents per kilowatt hour (kWh) for landfill gas to 80.2 cents per kWh for solar photovoltaics (PV) in 2009.[1] Unlike the German model, but in line with many global FIT programs, solar and wind power rates are contingent on achieving high levels of domestic content in goods, services, or labour. In 2009, Ontario's LCRs started at 25 percent for wind and 50 percent for solar projects that were larger than 10 kilowatts (kW). These grew to 50 percent for wind and 60 percent for solar by 2012. LCRs for MicroFIT projects under 10 kW were set at 40 percent and rose to 60 percent by 2012.[2]

As a result of the policy shift, Ontario is now Canada's leading province in solar PV capacity, with over 700 megawatts (MW) online (Ontario Ministry of Energy 2013a, 5). In 2012, wind energy generated 3 percent of Ontario's electricity (6), while solar, wind, and bioenergy sources currently account for 3,000 MW of electrical generation capacity (61). This was a remarkable change from 2008, when there were only 2 MW of installed solar PV and about 700 MW of installed wind in Ontario's overall power mix. The government's most recent Long-Term Energy Plan (LTEP) expects 20,000 MW of renewable energy to be online by 2025. The phase-in period for wind, solar, and bioenergy has been extended to 2021, with an expected 10,700 MW online (Ontario Ministry of Energy 2013a). Procurement targets of 200 MW a year through 2018 are already in place for smaller-scale FIT projects. A new competitive Request for Proposal (RFP) process, which will provide for greater community involvement in locating projects, is currently being developed for large-scale projects (ibid.).

A precise account of the new jobs created from the GEA is not available as a result of poor recordkeeping compared with other jurisdictions that have implemented LCRs (Winfield 2013). But this does not mean that evidence of job creation does not exist. The Ontario government estimates that 20,000 jobs have been created in manufacturing and associated services to maintain the new renewable power projects. Over the past four years, labour and environmental groups that support the act have periodically listed new manufacturing sites of solar, wind, and other renewable components (Blue Green Canada 2011). And companies that at first openly opposed the LCRs, such as Vestas of Spain and General Electric (GE), ended up signing FIT contracts with the Ontario government. Likewise, in Quebec, GE, Enercon, and Repower opened manufacturing facilities in order to meet that province's LCRs on green power procurements (Kuntze and Moerenhout 2013, 22–23). In the Canada Renewables case, the WTO dispute panel stated plainly that the evidence it considered "discloses that the FIT Programme has been a key factor motivating a number of manufacturers to establish facilities for the production of renewable energy equipment in Ontario" (WTO 2013, 62).

International experience with LCR-dependent FIT schemes also suggests a positive correlation with employment. Medium- and longer-term job creation in engineering, technical support, and other green service sectors should be more or less permanent wherever electricity grids transition to renewable power. In fact, a 2012 study by a coalition of Canadian environmental and labour groups suggests that if current global wind power targets are met, global employment in this sector alone will more than double, from 670,000 workers in 2012 to over 1.4 million by 2030 (Blue Green Canada 2012). The Ontario government target was 50,000 jobs from all renewable sectors, with an auditor general's report suggesting that 40,000 of those could be directly related to the renewable energy sector and that "30,000, or 75%, of these jobs would be construction jobs and would last only from one to three years, while the remaining 10,000 would be long-term jobs in manufacturing, operations, maintenance, and engineering" (Auditor General of Ontario 2011, 117).

These numbers would need to be adjusted to take into account Ontario's missed jobs target, but they suggest that investment in renewables can help, to some extent, offset the general decline in local manufacturing. When the GEA was enacted, the auto industry, the engine of the province's manufacturing sector, had been particularly hard hit, losing 12,000 jobs in 2008 and 2009 alone.[3] As an official who helped draft the GEA explained,

> At the time, the global financial crisis was unravelling and GM and other automobile makers, which are big employers in the province, were shedding jobs. All in all, these events highlighted the need to find new sources of jobs and economic growth for the province. One obvious route was the continuing development of our renewable energy sector. (*Green Energy Reporter* 2010)

Opponents of localization schemes point to the higher consumer costs and global supply chain inefficiencies created by LCRs. Their arguments tend to be based on overly simple economic models and, in Ontario, exaggerated claims about the relative weight of renewables in the increase in overall energy costs (Winfield 2013). Recognizing that green power generally creates a modest cost increase, one of the strongest arguments in favour of LCRs is that they are essential to bolster public support for policies to rapidly expand renewable energy production (Kuntze and Moerenhout 2013). Such expansion requires substantial up-front capital investment in order to achieve long-term economic and environmental benefits. State support in the form of high rates for renewable power would likewise be difficult to sustain unless there were obvious spinoffs in the form of new employment. Without broad public support, such investments are not politically feasible. (See Figure 1.1 below for examples of European renewable power projects that owe their success, in part, to LCRs.)

FIGURE 1.1
European LCR Success Stories

As the WTO dispute panel in the Canada–Renewable Energy case explained, and as experts generally agree, state support for renewable power production is essential for any country hoping to green its power grid. Public support for the high cost that this transition entails will naturally depend on spinoffs from FITs and other such policies. LCRs, and community involvement in power production, were essential to early European success with renewable power deployment.

For example, the Danish wind industry, one of the most successful in the world, was built through a combination of financial incentives and LCRs (Cho and Dubash 2003; Bolinger 2001; Danish Energy Agency 2009). In order to spur wind energy development, the Danish government required a utility to buy wind-generated energy at 85 percent of its net cost. These highly attractive rates were restricted to members of local co-operatives living close to the turbines. The policy was extraordinarily successful in increasing wind generation of electricity, while simultaneously encouraging local ownership and acceptance of an environmentally friendly technology.

Spain is another good example of an early mover on wind. Provincial LCRs were employed as early as 1994 in Galicia, Navarre, Castile and León, and Valencia, and these LCRs plus a federal FIT program are credited with helping the Spanish renewable energy firm Gamesa grow to become the second-largest wind turbine manufacturer in the world by 2002. Local content quotas were as high as 70 percent in two provinces, and the Navarre LCRs are said to have created 4,000 jobs (Kuntze and Moerenhout 2013). Until 2001, Gamesa was in a joint ownership with Danish company Vestas, and this technology transfer, or learning by doing, allowed the former to succeed much better when it eventually broke off on its own (ibid.).

Ironically, the local preference policies that Denmark and Spain used to launch their renewable energy industries and that helped inspire the Ontario policies are very similar to the ones challenged by the EU at the WTO.

In addition to encouraging the production of renewable energy, the GEA is also intended to foster "a culture of conservation by assisting homeowners, government, schools and industrial employers to transition to lower energy use" (Ontario Ministry of Energy 2009). Conservation and energy efficiency measures are "generally accepted as the least expensive, lowest impact form of meeting new energy demand" (Calvert and Lee 2012, 8). They are also a recognized source of new green job creation – both induced by and directly associated with, for example, making buildings, electrical systems, and other infrastructure more energy efficient. One US estimate, based on a methodology developed by Max Wei, Shana Patadia, and Daniel M. Kammena (2010), suggests that efficiency and conservation creates 9 direct and 5.2 indirect jobs per $1 million invested.

Progress on the energy conservation front has been slow. Implementing the four principal conservation commitments – ambitious energy efficiency standards, mandatory home energy audits before sale, improvements to Ontario's building code, and greening the broader public sector – has been fragmentary and limited. As the Environmental Commissioner of Ontario stressed in a 2011 report, maintaining a balance between developing new sources of clean energy and implementing conservation measures is vital. In a welcome move, the province's energy minister committed again in mid-2013 "to investing in conservation before new generation, wherever that's cost-effective" (Spears 2013).

On balance, the benefits in the form of new job creation and lower emissions from phasing out coal plants appear to have already justified the slightly higher rates that Ontarians are paying for electricity. Unfortunately, the recent WTO decision declaring that the policy, specifically its job-creating LCRs, violates basic non-discrimination rules in the GATT and the TRIMs Agreement threatens to significantly undermine its economic development potential. (See Figure 1.2 below for details on other outstanding disputes.) We will now turn to that dispute before considering ways in which Ontario, and other jurisdictions, might salvage their policy space in order to capture more of the employment benefits from deploying green energy.

THE WTO DISPUTE

In September 2010, barely a year into the GEA's new life, Japan launched a complaint at the WTO regarding the LCRs of the act's FIT regime. In its subsequent request to establish a dispute panel, filed with the WTO in June 2011, the Japanese government explained that the measures "are inconsistent with Canada's obligations under the *SCM [Subsidies and Countervailing Measures] Agreement*, the GATT 1994, and the *TRIMs Agreement* because they constitute a prohibited subsidy, and also discriminate against equipment for renewable energy generation facilities produced outside Ontario" (WTO 2011, 3). The three specific violations listed were:

1. Articles 3.1(b) and 3.2 of the *SCM Agreement*, because the measures are subsidies within the meaning of Article 1.1 of the *SCM Agreement* that are provided contingent upon the use of domestic over imported goods, namely contingent upon the use of equipment for renewable energy generation facilities produced in Ontario over such equipment imported from other WTO Members such as Japan;
2. Article III:4 of the GATT 1994, because the measures accord less favourable treatment to imported equipment for renewable energy generation facilities than accorded to like products originating in Ontario; and
3. Article 2.1 of the *TRIMs Agreement*, in conjunction with paragraph 1(a) of the Agreement's Illustrative List, because the measures are trade-related

FIGURE 1.2
WTO Loss, AIT, and NAFTA Disputes Outstanding

In addition to the WTO dispute, the GEA and associated decisions were simultaneously challenged through the North American Free Trade Agreement (NAFTA), under its investor-to-state dispute settlement provisions, and the Agreement on Internal Trade (AIT).

NAFTA: Mesa Power, a US investor with holdings in Alberta, disputes what it considers the arbitrariness of Ontario's approvals process for new FIT projects that were proposed in the run-up to that province's 2011 provincial election. In August that year, just three months after submitting a notice of intent to file for NAFTA Chapter 11 arbitration, Mesa filed its notice of arbitration, which claimed violations of national treatment, most favoured nation treatment, and fair and equitable treatment (Mesa Power Group 2011). The company further claimed a violation of NAFTA Article 1106, which prohibits performance requirements, including the use of domestic content and the conferring of a benefit in exchange for the use of domestic content. Although the NAFTA procurement chapter applies only to limited federal government agencies, Mesa claims that the Ontario policy is covered by the broader investment chapter. The WTO Dispute Settlement Body and Appellate Body decisions support this view and may factor into the Chapter 11 tribunal decision. The company is seeking $775 million in damages.

AIT: In September 2011, the governments of Alberta, British Columbia, and Saskatchewan jointly requested consultations with Ontario on its FIT program, which they claimed was discriminatory to investors because of its LCRs. According to a summary of AIT cases in October 2014, consultations in the case continue (AIT 2014). Like NAFTA's Article 1106, the AIT's Chapter 6 on investment, where the dispute lies in this case, includes a prohibition on performance requirements, including to achieve a specific level or percentage of local content of goods or services, purchase or use goods or services produced locally, or purchase goods or services from a local source (AIT 2012). However, the AIT does not apply to energy goods or services until the provinces and territories can agree to include rules for energy in an unfinished Chapter 12. The dispute is even more curious considering that a clause in the more liberalizing New West Partnership Trade Agreement immunizes renewable energy support policies from interprovincial trade or investment disputes (British Columbia, Alberta, and Saskatchewan 2009).

> investment measures inconsistent with Article III:4 of the GATT 1994 which require the purchase or use by enterprises of equipment for renewable energy generation facilities of Ontario origin. (WTO 2011, 3)

A WTO dispute panel was established in July 2011. The EU initiated an almost identical complaint against Canada in August 2011, and a second dispute settlement panel was established in January 2012. The disputes proceeded in parallel from that point, and on 19 December 2012, the two WTO panels jointly ruled against Canada on the GATT and the TRIMs Agreement charges, but they did not agree with Japan or the EU that the FIT was an illegal subsidy. In May 2013, the WTO Appellate Body confirmed this result, although with significantly different legal reasoning.

Dispute Panel Decision

The core question in the Canada Renewables dispute, and the issue on which Canada's entire defence rested, was whether the FIT and associated LCRs were exempted from the national treatment obligations of the GATT by virtue of Article III:8(a) (WTO 2012a, para. 7.113). The article specifies that the national treatment (non-discrimination) obligation "shall not apply to laws, regulations or requirements governing the procurement by governmental agencies of products purchased for governmental purposes and not with a view to commercial resale or with a view to use in the production of goods for commercial sale" (WTO 1995, 491). This exclusion was intended to preserve the ability of GATT parties and WTO member governments to use government procurement as a policy tool without running afoul of GATT obligations.

As the WTO panel noted, however, Article III:8(a) stipulates several conditions that must be met for a measure governing procurement to fall outside the scope of the GATT national treatment obligation. Only if a measure satisfies all these conditions would the national treatment obligation not apply. As broken down by the panel, the three pertinent questions were:

> (i) whether the challenged measures can be characterized as "laws, regulations or requirements governing procurement";
> (ii) whether the challenged measures involve "*procurement* by governmental agencies"; and
> (iii) whether any "procurement" that exists is undertaken "for *governmental purposes* and not with a view to *commercial resale* or with a view to use in the production of goods for commercial sale." (WTO 2012a, para. 7.122)

The panel explicitly rejected EU arguments that to qualify as a measure governing procurement, the minimum domestic content must directly relate to the product procured by the government (in this case, electricity). Accordingly, it concluded that even though the LCRs apply to generation equipment and not the product directly purchased (electricity), they "should be properly characterized as one of the 'requirements governing' the alleged procurement of electricity for the purpose of Article III:8(a)" (WTO 2012a, para. 7.128). The panel also rejected Japan's argument that government procurement must involve governmental use, consumption, or benefit of the procured product. If Japan's narrow characterization had been accepted, then the purchase of electricity, which is consumed by the population at large, could not be defined as government procurement. Instead, the panel interpreted *procurement* to have the same, ordinary meaning as *purchase*. In the panel's view, the purchase of electricity by the Ontario government (to meet the needs of its population) constituted procurement within the meaning of Article III:8(a) (WTO 2012a, para. 7.131).

Ultimately, the WTO panel's ruling hinged on the third element – specifically, the requirement that an excluded procurement must not be made with a view to commercial resale. Canada had argued that commercial resale "means a purchase with the aim to resell for profit" (WTO 2012a, para. 7.146). The OPA, which procures the electricity, is not a commercial, profit-seeking entity and is in fact barred by legislation from profiting by its purchases of electricity (WTO 2012b, A-65, para. 31). The OPA's primary purpose in entering into long-term contracts with electricity suppliers is to ensure sufficient investment in additional capacity to meet Ontario's electricity needs. Indeed, the OPA was created in 2004 as a public policy response to the failure of the partially liberalized market to yield significant new private investment in electricity generation.

The panel did not contest Canada's factual arguments that the OPA does not seek to profit from the purchase of renewable electricity and simply recovers its costs when it supplies the electricity to the grid. But it decided that because Hydro One, a provincial government entity, and municipal public utilities ultimately resell the renewable electricity procured by the OPA, "the Government of Ontario and the municipal governments not only profit from the resale of electricity that is purchased under the FIT Programme, but also that electricity resales are made in competition with licensed electricity retailers" (WTO 2012a, para. 7.151). The panel therefore concluded that the procurement of electricity under the FIT program is undertaken with a view to commercial resale. Since all three GATT Article III:4 conditions were not met, the panel decided that the FIT scheme with its LCRs was not an exempted form of government procurement. It followed inexorably that the LCRs associated with the FIT contracts for renewable energy were trade-related investment measures, which, by favouring local over imported products, were inconsistent with Article III:4 of the GATT 1994 and with Article 2.1 of the TRIMs Agreement.

Appellate Body Review

Canada appealed the panel's decisions to the WTO Appellate Body, insisting again that Ontario's measures were an excluded public procurement. But on 6 May 2013, the Appellate Body upheld the panel's ruling that the FIT program's LCRs were inconsistent with Canada's obligations under the GATT 1994 and the TRIMs Agreement (WTO 2013). However, its legal reasoning differed from that of the panel, overruling it on key interpretive points. Notably, the Appellate Body adopted a reading that considerably narrowed the scope of procurement-related measures safeguarded from challenge by Article III:8(a). In the panel decision, the Canadian defence was tripped up on the third element of Article III:8(a) regarding commercial resale; in the Appellate Body ruling, however, the contested measures did not even clear the first hurdle of the exclusion.

In its view, the FIT contracts and associated LCRs could not be characterized as "laws, regulations, or requirements governing procurement" within the meaning of Article III:8(a) because that article "does not cover discriminatory treatment of the equipment used to generate the electricity that is procured by the Government of Ontario" (WTO 2013, para. 5.84). The Appellate Body jurists attached great significance to a distinction in Article III:8(a) between procurement and products purchased, a matter considered of little importance by the panel.[4] The Appellate Body asserted that the difference in wording was deliberate, with procurement referring, in its view, to "the process by which government purchases products" and a product, understood within the broader context of Article III, referring to "something that is capable of being traded" (WTO 2013, para. 5.62).

In a complicated chain of reasoning, the Appellate Body judged that "the derogation of Article III:8(a) must be understood in relation to the obligations stipulated in Article III" as a whole (WTO 2013, para. 5.3.3). It then noted that other paragraphs of Article III prohibit treating imported products less favourably than like, directly competitive, or substitutable products of national origin. The Appellate Body imported these attributes into the simple reference to products in Article III.8(a). Finally, it asserted that in order for a measure to be shielded by the government procurement exclusion, the foreign products allegedly discriminated against must be in a competitive relationship with the products purchased by governmental agencies.

The panel had found that the OPA's purchases of electricity fell, in principle,

> within the scope of the derogation of Article III:8(a), because the generation equipment "is needed and used" to produce the electricity, and therefore there is a "close relationship" between the products affected by the domestic content requirements (generation equipment) and the product procured (electricity). (WTO 2013, para. 5.76)

But, relying on its freshly minted criterion of competitive relationship, the Appellate Body reversed the panel's finding that "the 'Minimum Required Domestic Content Level' should be properly characterized as one of the 'requirements governing' the alleged procurement of electricity for the purpose of Article III:8(a)" (WTO 2012a, para. 7.128). In the view of the Appellate Body, because the Ontario measures applied to generation equipment, which was not in a competitive relationship with the product purchased (that is, electricity), "the Minimum Required Domestic Content Levels cannot be characterized as 'laws, regulations or requirements governing the procurement by governmental agencies' of electricity within the meaning of Article III:8(a) of the GATT 1994" (WTO 2013, para. 5.79).

Global Fallout from the WTO Renewables Decision

The Canada Renewables dispute was the first case in the history of the GATT and the WTO in which the dispute settlement authorities were asked to interpret the procurement exclusion in Article III:8(a).[5] That article was assumed by most legal experts and governments to "expressly exclude government procurement from the GATT national treatment obligations" (Johnson 1998, 202; see also Arrowsmith 2011). Perhaps reflecting the times, including strong pressure from the EU and United States to expand coverage of government procurement at the WTO and in bilateral and regional trade liberalization treaties, the WTO decision has restricted the flexibility to apply preferential purchasing policies that many member governments understood they had preserved by excluding government procurement from national treatment. In a clear instance of ignoring the forest for the trees, the Appellate Body painstakingly parsed each term in the exclusion, considerably narrowing its protective scope and thereby frustrating the clear sense of the provision as a whole.

In Ontario, the distressing result is that the ruling further undermined support for one of the most innovative green energy policies in North America during a period of rapidly rising GHG emissions and dangerous global climate change. At the same time, the result emboldened other countries to challenge similar FIT programs in other parts of the world. China has filed a dispute against LCRs in renewable energy programs in Italy and Greece. Likewise, the United States is challenging India's Jawaharlal Nehru National Solar Mission (JNNSM), which introduced an LCR in 2010 requiring developers to source crystalline silicon (CSi) modules in India (see Table 1.1 below).

Given the number of trade disputes involving renewable energy, the International Centre for Trade and Sustainable Development proposes the need for a Sustainable Energy Trade Agreement, and it encourages WTO member governments to direct disputes involving LCRs away from the renewable energy sector in recognition of their use in shifting countries toward more green power production. Given the stalemate in the Doha

TABLE 1.1
LCRs in Dispute

Country/ province	*Type of LCR*	*Estimated impact*	*Disputes*
Ontario	FIT requirement 50% LCR for solar and 60% LCR for wind.	Government estimates 20,000 jobs created; modest power price increases; 20,000 MW of renewable energy generation by 2025, representing half of all capacity; amount of new generation; coal plants closed by 2014.	WTO, NAFTA (Chapter 11), AIT.
Quebec	RFP with between 30% and 60% LCR allocated to specific regions – focus on manufacturing capacity.	Not yet determined.	None yet.
India	JNNSM introduced LCR in 2010, by which solar developers must purchase locally manufactured CSi modules (exempts thin film technology).	LCR focused on manufacturing rather than services and other downstream benefits. Fully 70% of developers favour thin film to avoid the LCR, despite better efficiency of CSi, partly because of US low-interest loans to Indian firms purchasing thin film from US suppliers. Still, Peterson Institute estimates 3% to 7% additional growth in CSi market because of LCR.	US formally brought JNNSM to WTO dispute (DS456) in February 2013. India claims it should receive differential treatment as a developing country and that LCR is procurement by the public National Thermal Power Corporation. Dispute panel established 23 May 2014.
China	Starting in 1997, RFPs with LCR, then FITs with LCR up to 2009. Some LCRs phased out after first US WTO challenge. Clean Development Mechanism also provided financing for projects that were Chinese owned or joint ventures with Chinese companies.	Very successful: created high output and successful global export players. From 1,260 MW installed wind capacity in 2005 to 25,805 MW at end of 2009. Estimated 150,000–200,000 jobs created by growth in wind capacity.	US requested WTO consultations (DS419) in December 2010, claiming that Chinese funds or awards to enterprises manufacturing wind power equipment contingent on LCR violated the Agreement on Subsidies and Countervailing Measures.

TABLE 1.1 (continued)

Country/ province	*Type of LCR*	*Estimated impact*	*Disputes*
Italy, France, Greece	In 2011, Italy introduced 5–10% LCRs (sourced from the EU) to subsidize solar (Conto Energia). In France, 10% bonus on the price the EDF (electricity utility) pays for solar as long as there is a 60% EU-wide LCR (decreased to 30% after one year due to cost).	9 GW installed solar PV in Italy by end of 2011, more than China or Germany – a tripling from the previous year.	China filed WTO dispute (DS452) in November 2012 against LCRs in several EU countries, including Italy and Greece.
Jordan, Saudi Arabia	Jordan FIT introduced in December 2012 with 15% LCR. Saudi Arabia will introduce LCR in second round of bidding on 2,850 MW of renewable energy.	Not yet determined.	None.
United States	Financial incentives for advanced energy storage from California suppliers; incentives to homeowners in Massachusetts who use state-made solar PV; incentive energy rates for wind turbine installations using state-manufactured turbines.	Not yet determined.	China disputing LCRs in the renewable energy policies of several US states, while challenging US countervailing measures on what it considers China's ongoing subsidization of solar.
Brazil	Has applied an LCR of 60% for wind energy to encourage manufacturing; this is required to access loans from the National Development Bank.	Encouraged Gamesa (Spain) and Sinovel (China) to set up manufacturing in Brazil.	None.

Source: Authors' compilation, mainly from Kuntze and Moerenhout (2013).

Round of trade negotiations at the WTO, let alone in international climate negotiations, however, such a negotiated solution cannot possibly succeed if the goal is to quickly phase out polluting power sources. A much quicker and surely less controversial option for national and sub-national governments would involve a more proactive approach to generating green power. In markets the size of Ontario, there could be more positive job creation in a public sector solution involving not only direct procurement of energy projects by provincial and local utilities but also added emphasis on downstream employment in services.

Compliance but Not Complacency with the WTO

On 20 June 2013, Canada formally notified the WTO that the government of Ontario intended to implement its ruling in the renewables dispute. In August 2013, the OPA published a directive from the provincial government lowering the LCRs for all future FIT and MicroFIT projects to:

- 20 percent for onshore wind facilities
- 22 percent for solar PV using CSi technology
- 28 percent for solar PV using thin-film technology
- 19 percent for solar PV using concentrated PV technology

In early 2014, the minority Liberal government introduced legislation to eliminate LCRs from all future FITs under the GEA, and this change came into effect on 25 July 2014 (SunRise Power 2014) under the newly elected Liberal majority government. This should mostly satisfy Japan and the EU, as well as the US government, which submitted extensive argumentation against the Ontario policy to the WTO panel. The US government also prioritizes the elimination of so-called localization barriers to trade in its current international trade negotiations, albeit with an emphasis on the Asia-Pacific region (USTR 2013). While Ontario labour and civil society groups have urged the province not to change the LCRs, industry associations and some environmental groups argue that the policy has already achieved its objective of creating a new green energy market in Ontario, with future growth opportunities (Hall 2013).

LCRs are desirable for their potential to draw new manufacturing to Ontario. While recognizing that the empirical data is lacking with respect to the number of new manufacturing, technology services, and renewable energy development enterprises (Winfield 2013), and that the government's 20,000 jobs figure is based largely on econometric studies, there is nonetheless a strong indication that sustained investment in renewable power has and will continue to contribute to temporary and long-term new employment. A shift from incentivizing private sector–led growth to public sector–managed renewable power planning would provide even more employment and climate benefits to the province, while shielding LCRs from future trade challenges.

While the legitimacy of the Appellate Body's narrow interpretation of the government procurement exclusion is highly questionable, it is also true that the complexity of Ontario's partially liberalized electricity system unnecessarily exposed the economic development aspects of the GEA to challenge (Walkom 2012). The most straightforward option for preserving the local economic development component of the GEA, and to ensure that it is implemented consistently under current international trade rules, is for Ontario to pursue its complementary renewable energy and economic development goals through more conventional public procurement models. In the case of renewable energy contracts, where a public entity acquires the generation equipment, the Ontario government would still be free under WTO rules and Ontario's current WTO obligations to stipulate that all or any portion of that equipment be manufactured in Ontario.

The key to meeting the stringent criteria imposed by the Appellate Body is to apply the local economic development criteria to the generation equipment procured by a government entity, which then supplies the electricity using these generation resources. Reselling the electricity should not be an issue because the local development criteria apply directly to the purchased generation equipment, not to the electricity that may be resold.

While removing the LCRs, the Ontario government has also recently eliminated the larger FIT program and will be replacing it with an RFP similar to Quebec's. Ontario Power Generation (OPG) will be able to compete in these RFPs (Ontario Ministry of Energy 2013b), and, where it is successful, it would be completely free, under Canada's and Ontario's current WTO obligations, to require that the goods, labour, and services it needs to supply the contracted electricity contain prescribed levels of local content.[6] Alternatively, the OPG could simply be mandated by the Ontario government to supply increased amounts of renewable energy and to apply LCRs in its purchasing of the goods, services, and construction required to generate that electricity. This return to a more traditional, public sector energy procurement model would also be fully consistent with Canada's and Ontario's current obligations under the WTO's Agreement on Government Procurement.

In another promising reform, which was incorporated into the 2013 LTEP, the provincial government will encourage public sector entities to generate more renewable energy. In fact, while the earlier FITs were designed to encourage smaller community-based and Aboriginal or farm-based production, in practice the ability of these groups to finance projects was limited, and, as a result, "the Ontario program ended up being dominated by large commercial scale developers, who did not require such high rates for their projects to be viable" (Winfield 2013, 23). Reforms to the GEA include participation points, which give broader public sector entities – including municipalities, universities, schools, and hospitals – an advantage in tendering decisions; "price adders," which top up standard FIT payments for publicly owned projects; capacity set-asides, which direct the OPA to reserve capacity (initially 24 MW) for public sector projects; and direct financial support to assist municipalities and other public sector entities in the development and design of project proposals (Fyfe and Corpuz 2013).

Under Canada's current international trade treaty obligations, a renewable energy project owned by a municipality or a broader public sector entity remains free to apply LCRs in its purchases. For example, so long as it retains ownership of the generating equipment, a hospital or university developing rooftop solar panel systems under the MicroFIT program or new competitive RFP could apply preferences for local content in the supply of components. Consistent with the WTO, the provincial government could mandate that these broader public sector entities maximize local content as a condition for eligibility for premium rates and other advantages. Studies indicate that local ownership provides greater long-term economic development benefits.[7]

There are clear but understated cost benefits to public ownership of energy generation and transmission, the most obvious being that there is no required rate of return on investment. In the case of green energy – wind, solar, biomass, etc. – where even the WTO recognizes that incentives are necessary to ensure an adequate, stable supply,[8] the public is paying higher prices already through FITs that would not be required under a traditional public sector RFP. In Germany, local and state governments are re-municipalizing many electricity concessions, almost all of which are up for renewal before 2016, because of these benefits. "Between 2007 and mid-2012, over 60 new local public utilities (stadtwerke) have been set up and more than 190 concessions for energy distribution networks – the great majority being electricity distribution networks – have returned to public hands," explains a Public Services International briefing on renewable energy re-municipalization (Hall 2012, 10). The German renewables plan is extensive: the government hopes to phase out nuclear power and rely on renewables for 80 percent of its energy needs. It is estimated that this will require $20 billion per year up to 2050, with little economic incentive for the private sector to meet this challenge alone (Hall et al. 2013).

In summary, despite the WTO ruling in the Canada Renewables case, it remains feasible in Ontario and elsewhere to pursue LCRs and economic development goals effectively through traditional public sector procurement. Where provincial public entities are the purchasers and owners of the renewable energy generation equipment, they are free to apply local content criteria, and there are good reasons to continue to do so to a point.

Even by the Appellate Body's overly restrictive criteria, such purchases would fall within the scope of the exemption from national treatment.

SAVING THE PUBLIC OPTION: A FINAL CAUTION ON NEW TRADE NEGOTIATIONS

However, in the Canadian case, to continue implementing the economic development aims of the GEA, it will be essential to safeguard Ontario's existing policy flexibility over renewable energy procurement in ongoing trade and investment treaty negotiations led by the federal government – specifically, the Canada-EU Comprehensive Economic and Trade Agreement (CETA) negotiations that officially concluded in September 2014. Ideally, all provinces would want to do this, but a reading of the final text shows that only Ontario, Quebec, and Newfoundland and Labrador have made significant attempts to preserve their ability to apply LCRs on energy projects or to otherwise use energy or any other public procurement to achieve broader environmental or social objectives.

Procurement by provincially owned energy corporations (including Hydro One, OPG, and the OPA), municipal energy utilities, and broader public sector entities is not covered by Canada's existing international procurement obligations, including updated provincial coverage in the WTO Agreement on Government Procurement, which resulted from bilateral negotiations with the United States in 2010. These government entities are free to apply LCRs, or other local economic development criteria, in their procurement contracts for renewable energy generation equipment and related labour and services. Western provinces led by Alberta allege that Ontario's LCR violates non-discrimination rules in the AIT (see Figure 1.1 above), but the AIT clearly states that until an energy chapter (Chapter 12) is concluded, "no provision of this Agreement shall apply to any measure of a Party relating to energy goods or energy services as defined in Annex 1810.3" (AIT 2012, Art. 1810). Even the New West Partnership Trade Agreement among Canada's three westernmost provinces exempts "measures adopted or maintained to promote renewable and alternative energy" from its trade, investment, and procurement rules (British Columbia, Alberta, and Saskatchewan 2009, Part V, D2).

But the current federal government has an aggressive trade and investment treaty agenda and is presently negotiating over a dozen bilateral trade and investment agreements. The largest of these potential treaties involve the EU, Japan, India, and the 12-nation Trans-Pacific Partnership. The agreements threaten to further tie governments' hands in many areas only loosely related to trade, including sub-national government procurement.

The CETA is the closest of these negotiations to completion, with Canada and the EU announcing the conclusion of talks on 26 September 2014 and releasing a consolidated text, which must still undergo a legal "scrub" and translation into two dozen European languages. Before it comes fully into effect (in 2016 at the earliest), the CETA must be voted on by the European Parliament and ratified by all 28 EU member states. The investor-state dispute settlement provisions of the treaty are controversial in Europe, and its ratification is by no means assured.

As outlined in an October 2013 technical summary of the agreement in principle (Canada 2013), Canada's procurement offer to the EU covered virtually all sub-national government entities, including municipalities, academia, school boards, and hospitals (the MASH sector), Crown corporations, and other arm's-length government agencies, utilities, and construction projects by any of these entities. The Canadian technical summary explicitly referenced energy and transit – priority areas for the EU because of anticipated growth for EU-based infrastructure firms.

> - Coverage of 75–80 percent of procurement by major energy entities across Canada, with commitments by all provinces and territories with major energy-production and distribution capacity
> - Coverage of mass transit by all provinces and territories; Quebec and Ontario to retain a 25 percent Canadian value for the procurement of public-transit vehicles (rolling stock) with Quebec able to require that final assembly be in Canada within its 25 percent. (Canada 2013, 17)

Consequently, the CETA could dramatically reduce the existing policy flexibility in procurement in the renewable energy sector. Ontario, more than any other province, has wisely excluded several government entities from the agreement's procurement rules to preserve some room to prefer domestic goods and services in renewable energy production. For example, all "energy entities" except Hydro One and OPG are excluded. Furthermore,

> Ontario Power Generation reserves the right to accord a preference to tenders that provide benefits to the province, such as favouring local sub-contracting, in the context of procurements relating to the construction or maintenance of nuclear facilities or related services. A selection criterion of benefits to the province in the evaluation of tenders shall not exceed 20% of total points. (Foreign Affairs, Trade and Development Canada 2014)

Ontario's procurement commitments also exclude the provisions of the Ontario GEA, stating, "For greater certainty, nothing in this Agreement affects the procurement for the production, transmission and distribution of renewable energy, other than hydro-electricity, by the province of Ontario as set out in the *Green Energy Act*" (Foreign Affairs, Trade and Development Canada 2014).

In the wake of the Canada Renewables ruling, however, Ontario's procurement policies and associated LCRs are being restructured to comply with the WTO ruling. One of the most obvious steps, which the province has already begun to take, is to give a greater role to the OPG, which had been barred under the GEA from procuring renewable energy. Given this stated policy direction, covering the OPG under the procurement rules of the CETA, which Ontario has done for all procurement not related to the construction or maintenance of nuclear facilities, makes no sense. And exempting the GEA, as currently structured, will not safeguard Ontario's reformed policies from challenge.

Moreover, the recently announced price adders, participation points, capacity set-asides, and direct financial support for municipalities and broader public sector entities are all explicitly designed to provide these entities with a competitive advantage in the procurement process for renewable energy.[9] As such, they go against the grain of national treatment and other WTO and CETA procurement obligations. To fulfill the promise of this new policy direction, it would be prudent not to commit these entities under the procurement rules of the CETA. If such an unwise step were taken, the procurement of renewable energy, generating equipment, and associated goods, services, and construction would have to be fully and unequivocally excluded.

CONCLUSION

In an amicus curiae brief to the WTO panel, a coalition of Canadian unions and environmental groups asserted that the government of Ontario was providing leadership, notably absent at the federal level, in fulfilling Canada's binding commitments under the Kyoto Protocol to reduce GHG emissions. The brief also emphasized that a WTO ruling against the local economic development aspects of the GEA would contradict both the spirit and the letter of the United Nations' Framework Convention on Climate Change.[10]

Likewise, in their thorough study of LCRs in international renewables policy, and the new certainty that the WTO brings to the question of their legality under current international trade rules, Kuntze and Moerenhout stress,

> In spite of the legal reality ... a healthy debate concerning the usefulness or disadvantages of LCRs should include welfare effects such as increased employment – even if a certain level of protectionism would be necessary to reach such a goal. Since it is found in the mission statements of the WTO, World Bank, IMF, and ILO, employment can arguably be considered a global public good. (2013, 8)

The increasingly tangible threats posed by climate change necessitate making urgent, structural change to global energy systems. These threats require, at the very least, a deferential interpretation of existing trade treaty exclusions, one that provides maximum latitude for those governments intent on tackling the greatest environmental challenge of the era. Instead, however, the WTO Dispute Settlement Body, and especially the Appellate Body, have embraced an unacceptably narrow, restrictive interpretation of the GATT government procurement exclusion that not only undermines Ontario's efforts to pursue sustainable development but also casts a chill over similar initiatives around the world. This disappointing decision provides ample support for the view that the WTO is not a fit arbiter of green energy and environmental protection policies.

Despite this disappointing decision, it is still feasible to preserve the local economic development goals of the GEA and for public policy to sustain continuing growth and job creation in Ontario's renewable energy sector. The key to reconciling the GEA's sustainable development thrust with WTO obligations is to pursue these job creation goals through more traditional public sector procurement policies. There are cost benefits to doing so, and public support would likely be high, as it is in Europe. However, even this approach is threatened by the impending CETA, in which purchases of renewable energy by many provincial and local entities will be covered by international procurement restrictions for the first time, not to mention the 12-country Trans-Pacific Partnership. In light of the WTO decision, it will be even more essential for the Ontario government to fully safeguard its existing policy flexibility over procurement and renewable energy in ongoing trade and investment treaty negotiations directed by the federal government.

NOTES

1. FITs for solar and wind have decreased substantially since then. "The OPA achieved further cost savings with a significant reduction in the purchase price of renewable electricity in new FIT contracts. The lower FIT prices have reflected the reduction of domestic content requirements and a reduction in technology prices, saving $1.9 billion" (Ontario Ministry of Energy 2013a, 15).
2. A small FIT project is currently defined as a generating facility of between 10 kW and 250 kW (or up to 500 kW in the case of a facility connected to a line of 15 kilovolts or greater). A MicroFIT project is one with a capacity of less than 10 kW (Ontario Ministry of Energy 2013b, 2).
3. Canadian vehicle assembly declined by almost 20 percent between 2007 and 2010, resulting in the loss of 12,000 assembly jobs in 2008 and 2009 alone (CAW 2012, 3).
4. The panel concluded that "the term 'procurement,' when interpreted in its immediate context, should be understood to have the same meaning as the term 'purchase'" (WTO 2012a, para. 7.135).
5. "These proceedings are the first where a panel has been asked to interpret and apply Article III:8(a) of the GATT 1994" (WTO 2012a, para. 7.122).
6. The OPG is not a committed entity under the WTO's Agreement on Government Procurement.

7. "The National Renewable Energy Laboratory has verified that wind projects with 100% local ownership generate twice the long-term jobs and one to three times the economic impact of absentee owned wind projects" (Farrell 2011, 15); the National Renewable Energy Laboratory study is by Lantz and Tegen (2009).
8. "In the absence of demand that is more responsive (but not only for this reason), governments and regulators have sought to control potential/ actual price volatility by intervening in the market because of the value of stable electricity prices to their economies, with the consequence that many countries have experienced insufficient investment in generation because the price achieved on their 'organized' wholesale market is not allowed to rise to a level that, in the long-run, fully compensates generators for the all-in cost of their investments (including fixed and sunk costs). Private investors will not be willing to finance construction of new generation under such conditions; and in the absence of such investment, an electricity market will be unable to reliably meet future electricity demand. This is referred to as the 'missing money' problem, and it affects not only more expensive solar PV and wind generation technologies, but also 'conventional generating technologies, where energy-only markets do not support investment.' To resolve this dilemma, 'alternative mechanisms to wholesale spot markets have been required to provide incentives for long-term investment to meet forecasted demand,' including power purchase agreements (as in Ontario) and 'capacity' payments" (WTO 2013, para. 7.283; footnotes omitted).
9. As a legal analysis of these changes observes, "If the public sector entities structure their projects so that they have an equity interest in the projects, it will mean that their applications will be considered ahead of developer/third party applications. It will also mean that the competition for FIT contracts will initially be between public sector entities and then subsequently as between those public sector entities and private/third party developers. In short this will provide public sector entities with a competitive advantage in securing FIT Contracts over developer/third party applications" (Fyfe and Corpuz 2013, 3).
10. Article 3 of the Framework Convention reads, "The Parties have a right to, and should, promote sustainable development. Policies and measures to protect the climate system against human induced change should be appropriate for the specific conditions of each Party and should be integrated with national development programmes, taking into account that economic development is essential for adopting measures to address climate change" (quoted in Shrybman 2012, 9).

REFERENCES

AIT (Agreement on Internal Trade). 2012. *Agreement on Internal Trade. Consolidated Version*. Winnipeg, MB: Internal Trade Secretariat. http://www.ait-aci.ca/en/ait/ait_en.pdf.

—. 2014. "Summary of AIT Disputes (56)." http://www.ait-aci.ca/en/dispute/summary_en.pdf.

Arrowsmith, S., S. Treumer, J. Fejo, and L. Jiang. 2011. *Public Procurement Regulation: An Introduction*. Nottingham, UK: University of Nottingham. www.nottingham

.ac.uk/pprg/documentsarchive/asialinkmaterials/publicprocurement regulationintroduction.pdf.

Auditor General of Ontario. 2011. "Electricity Sector – Renewable Energy Initiatives." Chap. 3 Sec. 3.03 in *Annual Report*. http://www.auditor.on.ca/en/reports_en/en11/303en11.pdf.

Blue Green Canada. 2012. *More Bang for Our Buck: How Canada Can Create More Energy Jobs and Less Pollution*. Toronto: Blue Green Canada. http://bluegreen canada.ca/sites/default/files/resources/More%20Bang%20for%20Buck%20Nov%202012%20FINAL%20WEB.pdf.

Bolinger, M. 2001. *Community Wind Power Ownership Schemes in Europe and Their Relevance to the United States*. Berkeley, CA: Lawrence Berkeley National Laboratory.

British Columbia, Alberta, and Saskatchewan. 2009. *Canada's New West Partnership: New West Trade Partnership Agreement*. http://www.newwestpartnership trade.ca/pdf/NewWestPartnershipTradeAgreement.pdf.

Calvert, J., and M. Lee. 2012. *Clean Electricity, Conservation and Climate Justice in BC: Meeting Our Energy Needs in a Zero-Carbon Future*. Vancouver: Canadian Centre for Policy Alternatives.

Canada. 2013. "Technical Summary of Final Negotiated Outcomes: Canada–European Union Comprehensive Economic and Trade Agreement." Agreement-in-Principle. http://www.actionplan.gc.ca/sites/default/files/pdfs/ceta-technicalsummary.pdf.

CAW (Canadian Auto Workers). 2012. *Re-thinking Canada's Auto Industry: A Policy Vision to Escape the Race to the Bottom*. Toronto: CAW.

Cho, A.H., and N.K. Dubash. 2003. *Will Investment Rules Shrink Policy Space for Sustainable Development? Evidence from the Electricity Sector*. Washington, DC: World Resources Institute.

Danish Energy Agency. 2009. *Wind Turbines in Denmark*. Copenhagen: Danish Energy Agency. http://www.ens.dk/sites/ens.dk/files/dokumenter/publik ationer/downloads/wind_turbines_in_denmark.pdf.

Environmental Commissioner of Ontario. 2011. *Restoring Balance: A Review of the First Three Years of the Green Energy Act*. Vol. 1 of *Annual Energy Conservation Progress Report*. Toronto: Queen's Printer.

Farrell, J. 2011. "Maximizing Jobs from Clean Energy: Ontario's 'Buy Local' Energy Policy." Minneapolis, MN: Institute for Local Self-Reliance. http://ilsr.org/downloads/Maximizing+Jobs+From+Clean+Energy%3A+Ontario%E2%80%99s+%E2%80%98Buy+Local%E2%80%99+Policy/.

Foreign Affairs, Trade and Development Canada. 2014. "Consolidated CETA Text." Last modified 25 September. http://www.international.gc.ca/trade-agreements-accords-commerciaux/agr-acc/ceta-aecg/text-texte/21_01.aspx?lang=eng.

Fyfe, S.J., and B. Corpuz. 2013. "Green Revenue and Economic Development Opportunities for Public Sector Entities." Planning & Development Series. *Public Sector Digest* September. http://www.blg.com/en/NewsAndPublications/Documents/Green_Revenue_-_Public_Sector_Digest_-_SEP2013.pdf.

Green Energy Reporter. 2010. "Cornerstone Conversation: Pearl Ing, Ontario Energy Ministry." Interview, 26 October. http://greenenergyreporter.com/features/cornerstone-conversation-features/cornerstone-conversation-pearl-ing-director-renewable-energy-facilitation-office-ontario-energy-ministry/.

Hall, D. 2012. "Re-municipalising Municipal Services in Europe." Report commissioned by the European Federation of Public Service Unions. London: Public Services International Research Unit, Business School, University of Greenwich. http://www.right2water.eu/sites/water/files/re-municipalising%20municipal%20services%20in%20Europe%20-D.%20Hall%20May-Nov%202012.pdf.

—. 2013. "Trade Ruling May Not Hurt Ontario Wind Industry, Says Spokesman." *Windsor Star*, 7 May. http://blogs.windsorstar.com/2013/05/07/cwea-responds-to-wto-ruling/.

Hall, D., S. Thomas, S. van Niekerk, and J. Nguyen. 2013. "Renewable Energy Depends on the Public Not Private Sector." London: Public Services International Research Unit, Business School, University of Greenwich. http://www.psiru.org/reports/renewable-energy-depends-public-not-private-sector.

Johnson, J. 1998. *International Trade Law*. Toronto: Irwin Law.

Kuntze, C., and T. Moerenhout. 2013. *Local Content Requirements and the Renewable Energy Industry – A Good Match?* Geneva: International Centre for Trade and Sustainable Development.

Lantz, E., and S. Tegen. 2009. "Economic Development Impacts of Community Wind Projects: A Review and Empirical Evaluation." Paper NREL/CP-500-45555, to be presented at WINDPOWER 2009 Conference and Exhibition, Chicago, 4–7 May. http://www.nrel.gov/docs/fy09osti/45555.pdf.

McBride, S., and J. Shields. 2011. "Industrial Strategies for Green Jobs: Opportunities and Obstacles in the Ontario Case." Working paper. Toronto: Work in a Warming World Research Project, York University.

Mesa Power Group. 2011. "Notice of Arbitration under the Arbitration Rules of the United Nations Commission on International Trade Law and the North American Free Trade Agreement: Mesa Power Group, LLC v. Government of Canada." http://www.international.gc.ca/trade-agreements-accords-commerciaux/assets/pdfs/disp-diff/mesa-02.pdf.

Ontario. Ministry of Energy. 2009. "Ontario's Bold New Plan for a Green Economy." News release, 23 February. http://news.ontario.ca/mei/en/2009/02/ontarios-bold-new-plan-for-a-green-economy.html.

—. 2013a. *Achieving Balance: Ontario's Long-Term Energy Plan*. Toronto: Ministry of Energy. http://www.energy.gov.on.ca/en/files/2014/10/LTEP_2013_English_WEB.pdf.

—. 2013b. "New Direction to Ontario Power Authority Re: Renewable Energy." Toronto: Ministry of Energy.

Shrybman, S. 2012. "Amicus Submissions of Blue Green Canada; Canadian Auto Workers; Canadian Federation of Students; Canadian Union of Public Employees; Communications, Energy and Paperworkers Union of Canada; Council of Canadians; Ontario Public Service Employees Union." Toronto: Sack, Goldblatt Mitchell. http://canadians.org/sites/default/files/Trade/submission-WT-DS412-0512.pdf.

Spears, J. 2013. "New Nuclear Not Needed in Ontario, Green Groups Say." *Thestar.com*, 10 September. http://www.thestar.com/business/economy/2013/09/10/new_nuclear_not_needed_in_ontario_green_groups_say.html.

SunRise Power. 2014. "Ontario Reduces Solar PV FIT Rates and Scraps Local Content Rules." Blog post, 28 October. http://www.sunrisepower.ca/ontario-reduces-solar-pv-fit-rates-and-scraps-local-content-rules/.

USTR (Office of the United States Trade Representative). 2013."Localization Barriers to Trade." http://www.ustr.gov/trade-topics/localization-barriers.

Walkom, T. 2012. "Dalton McGuinty Scores an Assist as WTO Torpedoes Ontario Green Strategy." *Thestar.com*, 22 November. http://www.thestar.com/news/canada/2012/11/22/walkom_dalton_mcguinty_scores_an_assist_as_wto_torpedoes_ontario_green_strategy.html.

Wei, M., S. Patadia, and D.M. Kammena. 2010. "Putting Renewable and Energy Efficiency to Work: How Many Jobs Can the Clean Energy Industry Generate in the US?" *Energy Policy* 38:919–31.

Winfield, M. 2013. "Understanding the Economic Impact of Renewable Energy Initiatives: Assessing Ontario's Experience in a Comparative Context." Working paper. Toronto: Sustainable Energy Initiative, Faculty of Environmental Studies, York University.

WTO (World Trade Organization). 1995. "The General Agreement on Tariffs and Trade." Article III:8(a). In *The Results of the Uruguay Round of Multilateral Trade Negotiations: The Legal Texts*. Geneva: WTO.

—. 2011. "Canada – Certain Measures Affecting the Renewable Energy Generation Sector: Request for the Establishment of a Panel by Japan." Document 11-2786. Geneva: WTO. https://docs.wto.org/dol2fe/Pages/FE_Search/FE_S_S006.aspx?Query=%28%40Symbol%3d+wt%2fds412%2f*%29&Language=ENGLISH&Context=FomerScriptedSearch&languageUIChanged=true#.

—. 2012a. "Canada – Certain Measures Affecting the Renewable Energy Generation Sector; Canada – Measures Relating to the Feed-In Tariff Program: Reports of the Panels." Document 12-6849. Geneva: WTO. https://docs.wto.org/dol2fe/Pages/FE_Search/FE_S_S006.aspx?Query=%28%40Symbol%3d+wt%2fds412%2f*%29&Language=ENGLISH&Context=FomerScriptedSearch&languageUIChanged=true#.

—. 2012b. "Canada – Certain Measures Affecting the Renewable Energy Generation Sector; Canada – Measures Relating to the Feed-In Tariff Program: Reports of the Panels – Addendum." Document 12-6816. Geneva: WTO. https://docs.wto.org/dol2fe/Pages/FE_Search/FE_S_S006.aspx?Query=%28%40Symbol%3d+wt%2fds412%2f*%29&Language=ENGLISH&Context=FomerScriptedSearch&languageUIChanged=true#.

—. 2013. "Canada – Certain Measures Affecting the Renewable Energy Generation Sector. Dispute Settlement: Dispute DS412." http://www.wto.org/english/tratop_e/dispu_e/cases_e/ds412_e.htm.

CHAPTER 2

UNIONS AND CLIMATE CHANGE IN EUROPE: THE CONTRASTING EXPERIENCE OF GERMANY AND THE UNITED KINGDOM

JOHN CALVERT

INTRODUCTION

This chapter[1] examines how labour in Europe has addressed the issue of global warming, with a specific focus on the role of building trades workers and their unions in two countries: Germany and the United Kingdom. The built environment accounts for over 35 percent of greenhouse gas (GHG) emissions and close to 40 percent of energy use in Europe. Reducing the carbon footprint is thus key to European efforts to address global warming. This chapter reviews the overall climate policy framework of the European Union (EU) and examines the efforts of the two governments to green the built environment. It then looks, in more detail, at the specific initiatives taken by building trades unions to further the transition to a low-carbon economy and assesses the role of labour in shaping the restructuring of the respective construction industries to deal with global warming.

The ability of unions to play a constructive environmental role depends on the broader policy framework established by governments and on unions' influence in their own industry. Where union density is high and unions are significant players in training and workforce development, as in Germany, they have been influential in shaping the industry's environmental policies. In contrast, where unions are weak and have little

Work in a Warming World, edited by Carla Lipsig-Mummé and Stephen McBride. Kingston: School of Policy Studies, Queen's University.

say over training and workforce development, such as in the United Kingdom, they have had little impact in greening the industry. While the role of labour is only one factor in the success of climate policies, the presence – or absence – of union involvement does make a difference in the ability of countries to meet climate change objectives.

THE EUROPEAN PUBLIC POLICY CONTEXT

To understand the contribution of organized labour in construction to addressing climate change in Europe, it is necessary to locate its activities within the EU's overall environmental policy framework. This framework reflects a broad public consensus that global warming is real and that it poses a major threat (IPCC 1996, 2013; Stern 2006). Europeans recognize the urgent need to curb GHG emissions, lower energy use, and reduce reliance on fossil fuels. Accordingly, the European Parliament, European Commission (EC), and Council of Ministers legislated a package of measures in 2008, collectively known as the Climate and Energy Package, to reduce GHG emissions by 20 percent by 2020 from their 1991 levels, raise renewable energy's share to 20 percent, and improve energy efficiency by 20 percent. (The EU has since established a future objective of reducing GHG emissions by between 80 and 95 percent by 2050 compared to 1990.) Parliament ratified the 20-20-20 package in December 2008 (Morgera, Kulovesi, and Muñoz 2010; Hedegaard 2011).

Dissatisfied with the slow progress of member states, the EU issued another directive, number 2012/27/EU (25 October 2012), requiring additional measures. In January 2014, the EC proposed even more aggressive targets: a 40 percent reduction in GHG emissions and 27 percent of energy from renewables by 2030 (EC 2014).

A key issue for the EU has been how to upgrade workforce skills in light of predictions of shortages in major skills, which would impede the ability of member states to achieve their targets, thereby placing EU industries at a competitive international disadvantage (EMCO 2010b; Council of the European Union 2010, 2011). In construction, it has focused on upgrading workers' capacity for implementing new building technologies and renovating existing building stock (EMCO 2010a; EC 2011). It anticipates the emergence of new occupations for building renewable energy projects and low-carbon renovations. Its objective is to green all sectors of the EU's economy rather than focusing only on emerging green industries (ECORYS 2008).

In fact, a number of the EU's 27 member states have passed measures that exceed the EU's targets (Andrews and Karaisl 2010; Morgera, Kulovesi, and Muñoz 2010). These countries wish to become global leaders in climate mitigation and adaptation, and they view construction as critical to achieving their ambitious objectives.

The efforts of European trade unions, including unions representing workers in construction, have thus emerged within the context of this broader public consensus and public policy framework – a framework that has emphasized the urgent need for climate action and provided a range of public policy measures and targets to guide this process.

The EU has focused substantial attention on the building sector because buildings account for approximately 40 percent of GHG emissions and 35 percent of energy consumption. Measuring – and reducing – energy use is thus critical to implementing effective climate measures. As new buildings represent only about 1.5 percent of the existing stock, renovating existing structures is key to meeting the EU's climate objectives.

Labour's Role in the Development of the EU's Climate Policy

Organized labour is represented by the European Trade Union Confederation (ETUC), which represents 86 union federations in 38 European countries (not all in the EU). The ETUC supports tough climate targets (ETUC 2005a; Dupressoir et al. 2007; Kirton-Darling 2011) and endorses both the United Nations Framework Convention on Climate Change (UNFCCC) and the 1997 Kyoto Accord (ETUC 2005b, 2006, 2007, 2010; Decaillon 2009; Dupressoir et al. 2007). It recognizes the need to expand jobs in low-carbon industries, but it also believes that jobs in other, traditional sectors of the economy will disappear. There will be winners and losers. Consequently, it has a particular interest in how the labour process will be reorganized, arguing that workers and their unions must participate in shaping the future organization of work. It believes that sound labour market policy can facilitate the transition from declining to expanding sectors in a manner that minimizes disruption to workers' lives, protects their employment, and maintains their incomes (Decaillon 2009; Dupressoir et al. 2007; ETUC 2009; Kirton-Darling 2011). The ETUC refers to its approach as a just transition to a low-carbon economy.

Labour participates in numerous EU policy-making and consultative bodies, many of which include climate change in their mandate. While the EU has been incrementally expanding the scope and depth of the integrated European market – and this has a strong neoliberal focus – the quid pro quo of extending the reach of the market was to obtain labour's support, or at least acquiescence, to European integration by providing certain statutory rights for workers, such as those embodied in the 1989 Community Charter of the Fundamental Social Rights of Workers.

While labour has far less influence than business does on EU economic policy issues, it does play a modest role as a social partner in employment policy, labour relations, and training. For example, the EU's European Economic and Social Committee has an explicit mandate to involve representatives of the employed, as set out in the Treaty on the Functioning of the European Union (Article 300). EU unions have also used existing bipartite or tripartite institutions, including statutory works councils, to advance the climate interests of workers (Jagodzinski 2009). In a number of countries, unions also manage paritarian social funds, which provide unemployment insurance, pensions, sick leave, and other benefits. The extensive role of labour in the European climate debate thus reflects its broader involvement in EU policy-making (Laux 2008).

The European Construction Industry

The European construction industry is a major employer and a major driver of the economies of the EU's 28 member countries. In 2013, it employed 15.0 million, or 6.9 percent, of Europe's 216.8 million workers (OECD 2014). Construction accounted for 10.5 percent of Europe's gross domestic product (GDP). There are approximately 1.9 million employers. Most are very small, with 97 percent employing fewer than 20 workers (EFBWW 2014), but there are also some large, multinational construction contractors, active across Europe and around the world. Every EU member has a domestic construction industry, but there are significant variations in the organization of their industries and their labour market and training regimes.

Construction is different than many other industries. For example, it is strongly affected by fluctuations in the business cycle (Bosch and Philips 2003; Recio 2007), and demand for building work exhibits a boom-and-bust pattern more pronounced than in other economic sectors. In addition, the industry is disproportionately affected by seasonal factors, particularly in northern Europe, where work is more difficult in winter (Bosch and Philips 2003).

These factors increase the risks of construction. To minimize these risks, contractors transfer them to others (e.g., other firms, subcontractors, clients, project developers, purchasers of construction services, and building workers). Extensive subcontracting is a key component of risk shifting. Additionally, the EU has a significant underground construction economy, which reduces costs by exploiting undocumented workers, evading taxes, and ignoring regulations. These factors result in a highly competitive European construction labour market, which imposes downward pressure on costs while incessantly pushing firms to raise productivity (Bosch and Philips 2003).

A key challenge is workforce productivity. There are different ways to solve this problem. One is to focus on minimizing labour costs by organizing work in such a way that workers perform a limited number of simple, narrowly defined tasks – tasks organized and assigned by employers. This model can be traced back to F.W. Taylor's principles of scientific management (Braverman 1974). While requiring a core of qualified trades, it maximizes the employment of the least well trained – and hence cheapest – workers, normally through extensive subcontracting. It thus minimizes the investment made in the skills of the workforce (Bosch and Philips 2003; Brockmann, Clarke, and Winch 2008, 2010a, 2010b). With some exceptions, this is the approach that the UK industry has taken.

The alternative promotes productivity by supporting extensive training. It assumes that a highly qualified workforce will be a productive workforce. It requires lengthy apprenticeships, augmented by classroom study, to provide workers with the knowledge and skills they need for a lifelong career. It supports trainees financially through levies on employers, government grants, or social partnership funds. And it requires employers to provide jobs to apprentices to complete their training. Germany has followed the latter high-skill, high-wage approach.

The Role of Unions in the EU's Construction Industry

The European Federation of Building and Woodworkers (EFBWW) is the major union federation representing building workers. It is affiliated to the ETUC, bringing together 75 major construction unions from 31 European and neighbouring countries and representing over 2.3 million workers (EFBWW 2013). It represents the interests of its national affiliates in various EU-level forums and works with its employer counterpart, the European Construction Industry Federation (FIEC). The EFBWW's role parallels that of other sector-wide EU labour federations, such as the European Metalworkers' Federation and the European Transport Workers' Federation. It also works closely with the Building and Woodworkers' International (BWI), the key international construction union federation.

The EFBWW has worked with environmental non-governmental organizations (NGOs) to pressure the EU to implement Europe's climate targets more quickly. It also carries out a joint campaign with the Climate Action Network (CAN) Europe to press governments to accelerate the renovation of existing building stock (CAN and EFBWW 2011). CAN Europe is the largest environmental NGO coalition in Europe, representing 149 environmental organizations from 25 countries.

The EFBWW also coordinates 60 construction works councils established under EU Works Council Directive 94/45/EC. Their legal status gives them the right to participate in the management of their firms. Works councils (and the unions from which most members are selected) are another vehicle through which unions pursue climate objectives (ETUC and SDA 2009). The EFBWW also engages in a social partnership dialogue with the major construction employers in the EU, represented by FIEC (EFBWW and FIEC 2008, 2012).

The EFBWW and the BWI have drafted a Platform of Action for a Social and Green Europe, which includes 23 climate change demands for EU governments. Its proposals include making public investments in climate-friendly buildings and renovations, subsidizing environmentally sound infrastructure, providing additional green training for workers, and implementing energy conservation technologies throughout the built environment (EFBWW and BWI 2008; EFBWW and FIEC 2012).

The reason that training is such a major component of labour's climate agenda is that implementing low-carbon construction requires a high level of workforce skills and knowledge. Both new construction and retrofitting involve applying existing technologies to much higher tolerances as well as being able to work with new materials, equipment, technologies, and energy conservation systems. Low-carbon construction also requires workers to understand how their work contributes to the overall objective of reducing GHG emissions and energy consumption and, consequently, how it must be integrated into all of the other components of the construction process (Gleeson and Clarke 2013).

While the EFBWW articulates the collective voice of European labour, there is considerable variation in how much its national affiliates contribute to climate policies. Of particular concern for this study is how national differences are reflected in the economic and political influence of unions in Germany and the United Kingdom – and the corresponding role that unions play in shaping big-picture decisions associated with climate-related employment, training, and skills development. In the following sections, we will look at each country, focusing on the national climate policy framework, the role of national labour federations, and, finally, the role of the construction industry and its unions.

THE GERMAN PUBLIC POLICY RESPONSE TO CLIMATE CHANGE

German construction unions have long pursued climate initiatives within a supportive social and political context. For decades, Germany has been a leader in climate change mitigation, particularly through its support for renewable energy (Andrews and Karaisl 2010; Weidner and Mez 2008). In 1990, the federal government established a target of 25 percent carbon dioxide (CO_2) reduction by 2005, based on a 1987 benchmark (Watanabe and Mez 2004). In 1991, it introduced a feed-in tariff to stimulate the development of renewable electricity. In 2000, its National Climate Program aimed to reduce CO_2 emissions by between 50 and 70 million tonnes, and its *Renewable Energy Sources Act* targeted doubling the share of renewables in electricity consumption.

To address the estimated 37 percent of CO_2 emissions generated by buildings, the federal government introduced a Building Energy Efficiency Ordinance in 2002 to reduce energy use in new buildings by one-third (Weidner and Mez 2008). The ordinance established home insulation requirements for older homes and mandated that new homes have an energy certificate documenting their energy consumption. It also targeted replacing 2 million older furnaces within five years (Auer, Heymann, and Just 2008; Andrews and Karaisl 2010).

In 2007, the federal government introduced a new Integrated Energy and Climate Programme to support retrofitting existing buildings, and it raised the energy efficiency ratings for new buildings. Germany also committed to retrofitting 100,000 roofs with solar panels (Weidner and Mez 2008). Federal policies have been mirrored by state governments, some of whom, depending on resources and geography, have implemented more stringent programs (ibid.).

The German Construction Industry

Germany accounts for one-fifth of the EU's construction market, with construction activity totalling 260 billion Euros in 2012.[2] In 2013, it accounted for 56 percent of new investment and employed 2.8 million workers (OECD 2014), and in 2011, there were 391,000 companies (iXPOS 2014).

German construction is based on a strong craft tradition. "In marked contrast to other countries (e.g., the United Kingdom), the craft structure of the German construction labour market has … been strengthened in recent decades" (Bosch and Zühlke-Robinet 2003, 54). Apprenticeship expanded significantly between 1960 and 1980, with approximately 60 percent of school-leavers choosing careers through vocational education and training (VET) programs (Bosch and Charest 2008). It is a dual system, combining school-based and on-the-job training and involving the federal government, the *Länder* (federal states), social partners, and industry (Danish Technological Institute 2009). Despite pressures to deregulate the German labour market, its VET system remains intact, providing the basis for a highly skilled workforce of tradespeople.

Germany's construction labour market is highly regulated, with legislation specifying the craft skills that workers must possess to work in the industry (Bosch and Zühlke-Robinet 2003). The system imposes severe penalties – up to 100,000 Euros – on unqualified construction workers. There has been some growth in Germany's underground economy, but the regulations remain well enforced on larger projects.

Consequently, a majority of construction workers are well trained and expect a lifelong career. Productivity has risen rapidly in recent years (Bartholmai and Gornig 2007). According to a 1999 survey, 77 percent of building companies, employing 73 percent of construction workers, were registered as engaged in a handicraft trade. To be registered, these companies have to employ trained master crafts workers capable of supervising apprentices (Bosch and Zühlke-Robinet 2003). Very few German construction workers (5.6 percent) have no, or limited, qualifications; this contrasts with Britain, where the proportion of unskilled workers is estimated to be 33 percent (Richter 1998).

The Role of German Labour in Supporting Climate Change Initiatives

The *Deutscher Gewerkschaftsbund* (German Confederation of Trade Unions, or DGB) is the major union federation in Germany. It includes eight affiliated unions spanning the private sector and represents four-fifths of all unionized workers, including most organized construction workers (Addison, Schnabel, and Wagner 2006). However, union density in Germany has been declining, and only 18 percent of workers were union members in 2011 (OECD 2014). However, collective agreement coverage is much higher, at approximately 50 percent of the workforce in 2011 (54 percent western Germany, 37 percent eastern Germany); this is a result of legislation (Article 5 of the *Collective Agreement Act* and the *Posted Workers Act* in the construction sector) mandating that the terms of collective agreements apply to other workers in the same industry.

German workers have another vehicle to advance climate initiatives: statutory works councils. Established during the reconstruction of German industry (and the democratization of industrial relations following World War II), they give many workers the right to a voice in decisions relating to employment, training, and the organization of their workplaces (Behrens 2009).

Despite the decline in its membership over the past 25 years, the DGB remains a major force in shaping German labour market and industrial policy as well as a major player in the Social Democratic Party. It has strongly endorsed government climate policies, most notably in its 2008 New Deal on Environment, Economy and Employment. The DGB supports the government's Green New Deal, which has funded 30 major scientific projects to improve resource efficiency in German industry (Eurofound 2011).

The Alliance for Work and Environment also reflects the commitment of German unions on climate action. Established in 1998 by the DGB construction affiliate IG Bauen-Agrar-Umwelt (IG BAU), its objective was to create 200,000 new jobs through building conservation initiatives. The goal was to reduce CO_2 emissions by 2 million tonnes a year, lower the amount of energy used for heating by 80 percent, stimulate new energy conservation technologies, and reduce dependence on imported energy (UNEP 2006; Schneider 2009, 2010). Between 2001 and 2006, the government allocated 5.2 billion Euros to the project and retrofitted 342,000 apartments (UNEP 2009).

During Germany's 2007 presidency of the Council of the European Union, the DGB lobbied the German government to push for EU-wide building conservation standards. As a result, the EC issued an energy efficiency directive that year that applied to all EU member countries. The German unions worked closely with environmental NGOs and the ETUC on these initiatives (Schneider 2009).

In response to the 2008 recession, the DGB successfully lobbied the German government to include in its December 2008 stimulus package 4 billion Euros in low-interest loans to improve building efficiency. It also pressed for additional building sector investment in the subsequent stimulus package. In addition, Germany provided 8 billion Euros in stimulus money to renovate public sector buildings. The government has since claimed that its measures had renovated 617,000 apartments, created 292,000 jobs, and reduced CO_2 emissions by 1.4 million tonnes (Schneider 2010).

The Role of Unions in the German Construction Industry

The principal union representing German construction workers is IG BAU, which has about 306,000 members (Dribbusch and Birke 2012). IG BAU is affiliated through the DGB to the BWI and the EFBWW. Centrally coordinated negotiations on both the employer and the union sides result in binding collective agreements (with some regional flexibility in wages and salaries between western and former eastern Germany).

While the construction unions now represent less than two-fifths of the workforce, collective agreements establish prevailing employment standards throughout the industry. Fully 72 percent of construction workers were covered by these standards in 2007. German law gives the federal minister of labour authority to establish minimum wages for construction workers. In 2008, only 10 percent of construction workers earned less than two-thirds of the median construction wage (Bosch and Weinkopf 2010).

IG BAU has worked with Greenpeace and other environmental NGOs to promote investment in the energy-efficient refurbishment of buildings, including campaigns such as "Precedence of Jobs by Environmental Protection." Following the economic downturn of 2008, IG BAU pressured the federal and state governments to fund environmentally focused public infrastructure investments and prevent a major industry slump. These included additional subsidies for retrofitting buildings; green investments in schools, universities, child care centres, and municipal facilities; and loans to the construction industry to maintain investment levels (Laux, Schäfer, and Harnack 2010).

In sum, a number of interconnected factors have shaped the response of the German construction industry and its unions to climate change. Government policies, along with widespread public concern about global warming, have provided a supportive framework for advancing climate initiatives. Both the DGB and its construction affiliate IG BAU have been strong supporters of progressive climate policies. They have also recognized the potential benefits of investing in greening the built environment to both create jobs and address global warming.

In addition, the German construction workforce is highly trained and, consequently, well equipped to implement the numerous innovations in building technologies and systems needed to achieve aggressive climate targets. This high skill level also enables workers and the unions that represent them to contribute effectively both to the broader policy debate and to the hands-on development of Germany's climate initiatives.

THE UNITED KINGDOM'S PUBLIC POLICY RESPONSE TO CLIMATE CHANGE

As in the case of Germany, the United Kingdom's policy response to climate change has been shaped by both European and domestic factors. Global warming has been the subject of considerable public debate, encouraged by EU efforts, government initiatives, and the contribution of its scientific community. While public concern over global warming is, arguably, not as high as in Germany, there is nevertheless a widespread consensus that the problem is serious and that it requires significant government action.

The former Labour governments of Blair and Brown strongly supported international measures to limit global warming; however, the current Cameron-led coalition government has been backsliding on the issue.

The Labour government of Tony Blair commissioned London School of Economics economist Nicholas Stern to recommend the measures it should adopt (Stern 2006). Labour endorsed the EU's aggressive 20-20-20 targets (Parkes 2010; EC 2009). In 2008, it passed the *Climate Change Act*, which aimed to reduce CO_2 emissions by 26 percent by 2020 and 80 percent by 2050 compared with 1990. It also established a 30 percent renewable electricity target for 2020 and adopted the EU's Energy Performance of Buildings Directive (Billett and Bowerman 2009). All new public buildings are to be zero carbon by 2018 (in other words, produce no CO_2 emissions), and carbon emissions from the construction process are to be reduced by 15 percent by 2020 (UK 2008; UK Department of Energy and Climate Change 2009). The act established a Committee on Climate Change (CCC) to monitor progress. By 2013, the CCC had published five major reports to Parliament, assessing progress in reducing GHG emissions and energy use (UK CCC 2013a, 2013b).

The United Kingdom's built environment accounts for 36 percent of GHG emissions, according to the CCC; residential buildings represent two-thirds of this total. The government has established new zero-carbon standards to reduce energy consumption by 80 percent in new homes by 2016. The CCC estimates that total GHG emissions from buildings could be reduced by 74 percent by 2030 (UK CCC 2010). Since 2008, the United Kingdom has required all building owners, including landlords, to have an Energy Performance Certificate, which documents energy consumption.

The UK Construction Industry

With 2.9 million workers in 2011, construction is among the largest employers in the United Kingdom. It accounted for approximately £90 billion, or 6.7 percent, of GDP in 2011 and supported 280,000 businesses (UK BIS 2013). However, average company size is very small, with 93 percent of contractors employing fewer than 13 workers (Brockmann, Clarke, and Winch 2010a).

There is a broad consensus that the UK construction industry is poorly organized, inefficient, badly managed, and, in the view of some, chaotic (Bosch and Philips 2003; Brockman, Clarke, and Winch 2008; UK BIS 2013). Labour productivity is low compared with that in Germany. The industry relies heavily on subcontracting, much of which is concealed employment (Harvey 2001; Harvey and Behling 2008). The quality of training and apprenticeship has eroded in recent years, and a high proportion of UK building workers have limited skills. The country's ability to implement climate policy is thus compromised by the limitations of the industry itself. Efforts by the Conservative, New Labour, and coalition governments to address industry inefficiencies have focused largely on meeting the demands of building purchasers to cut costs and contractors to improve profitability rather than on upgrading workforce skills (Egan 1998, 2003).

The industry's limitations are reflected in its inability to deliver the government's climate objectives. This gap was dramatically highlighted by a 2007 Public Accounts Committee report that noted the absence of environmental monitoring on £3 billion worth of government-funded building projects. The committee found that "mandatory environmental assessments were carried out in only 35 percent of new builds and 18 percent of major refurbishment projects in 2005–06, and only 9 percent of projects could be shown to meet the required environmental standards" (UK Committee of Public Accounts 2007, 3).

Recognizing the weaknesses of the industry, the government commissioned a major study of low-carbon construction (UK BIS 2010a). The 2010 report noted that the industry was fragmented and poorly organized, with numerous separate silos rather than a coherent structure (UK BIS 2010b).

Poor workforce skills continue to undermine the United Kingdom's climate efforts. For example, the VET system is underfunded and poorly coordinated (Toner 2008; Clough 2009; Watson 2009). Government support is divided among a number of agencies. Colleges of further education provide classroom training, while Sector Skills Councils (SSCs) are responsible for assessing skills requirements and coordinating training among employers, colleges, and private training agencies (Brockmann, Clarke, and Winch 2010a). A separate Skills Funding Agency is responsible for training those over 18 years of age, but the Young People's Learning Agency deals with younger trainees. The delivery of training and apprenticeship has been shifted to colleges of further education, which now enrol two-thirds of trainees. But they have poor linkages to employers, and less than one-third of students now take traditional employer-sponsored apprenticeships.

UK construction training focuses heavily on narrow, specific skills that can be quickly learned rather than providing a comprehensive, well-rounded trades education. Employers avoid investing in workforce development except where it is to their immediate advantage. They want VET programs to give workers only the minimum skills (Brockmann, Clarke, and Winch 2008). Consequently, UK construction has a high proportion of low-wage, unskilled workers, approximately half of whom work on self-employment contracts and have no long-term attachment to the industry. Productivity is based on minimizing labour costs by keeping wages low rather than developing a highly trained workforce (Harvey 2001; Brockmann, Clarke, and Winch 2008; MacKenzie et al. 2010).

The Role of UK Labour in Addressing Climate Change

The United Kingdom's major union federation is the Trades Union Congress (TUC). With 6.5 million members, the TUC would appear to be a major political actor on climate issues. However, union membership has declined significantly since its peak in the late 1970s, when it approached 13 million members (TUC 2014). Union influence, both at the bargaining table and with government, has diminished accordingly. While much of this decline occurred under Margaret Thatcher, the later New Labour government was cool to union organizing, seeing labour market flexibility as a key competitive advantage. New Labour also resisted implementing some of the EU's labour rights directives.

Despite these constraints, the TUC has attempted to contribute to UK policy developments on climate change by publishing numerous research papers and policy statements as well as incorporating climate issues into its educational programs. As with other EU union federations, the TUC advocates a just transition framework to address climate change. This incorporates five key principles: sustainable development, inclusion of workers in policy-making, employment and income security during the transition, equitable sharing of costs, and a central role for government in shaping the transition rather than simply relying on market-based solutions (Dromey and Hunt 2009; TUC 2008).

While New Labour might have incorporated a labour-friendly approach into its climate program, it largely ignored the potential contribution of unions in shaping the United Kingdom's climate agenda. As a senior TUC representative noted, "Under New Labour, unions have had to adapt to a so-called 'post voluntary' system. It is very much about meeting employer demand through employer-dominated bodies, retaining employer prerogative, and not increasing collective bargaining over learning and skills" (Clough 2009, 14).

The new coalition government has been even less interested in a role for labour in climate-related training and skills development. Thus, the TUC has been largely marginalized in the United Kingdom's approach to climate change. More recently, its focus has shifted to dealing with the recession and the anti-labour policies of the Conservative-led coalition government. A review of its website, recent publications, and conference agendas indicates that other issues – such as unemployment, union organizing, training, employment rights, and workplace safety – have gained prominence since 2008.

The Role of Unions in the UK Construction Industry

The major construction unions are the Union of Construction, Allied Trades and Technicians (UCATT); UNITE, Britain's largest union, which has a construction division (formerly the Transport and General Workers' Union); and the General and Municipal Workers' Union. UCATT, with 84,377 members, is a construction union, while the other two are general unions, most of whose members work in areas other than construction (TUC 2014). Union membership is highest among the skilled trades and lowest among semi-skilled or unskilled manual construction workers, of which the United Kingdom has a much larger proportion than many other European countries. The Office for National Statistics estimates that union density in construction fell from 30.5 percent in 1995 to 14.5 percent in 2010 (Chamberlin 2011). In contrast to Germany, there is no legislation that extends the terms of collective agreements to unorganized workers.

UK employers have no commitment to social partnership and have successfully opposed the introduction of works councils. Low union density makes it difficult for unions to claim that they represent the average construction worker. Absent institutional arrangements giving them an effective voice in training, building unions are not able to act as effective advocates for the training needs of the industry, nor are they able to participate as partners with employers in shaping industry policies. In addition, they have witnessed a major decline in local government direct-labour schemes – an area where workforce training had been extensively implemented (UK BIS 2010c). A small membership also means that union resources are limited, making it difficult for unions like UCATT to participate effectively in industry education and training.

UK construction unions recognize the need to reskill the labour force to implement a low-carbon economy. They argue that many industry practices, such as self-employment, are major barriers to modernizing apprenticeship and training. And they have lobbied successive UK governments for reforms, largely without success. Despite these challenges, building unions have tried to push the SSCs to identify future skill requirements and develop programs to meet anticipated low-carbon training needs. UCATT participates in the Strategic Forum for Construction, a consultative organization composed of industry stakeholders, established in 2001 to engage in long-term strategic planning for the industry. UCATT also works with the National Skills Academy for Construction to promote sustainable construction practices (UCATT and ConstructionSkills Council 2006). However, opposition from unorganized employers, reluctant to see the union working with their employees, has undermined its efforts (UK BIS 2010c).

In sum, UK construction unions face major challenges in implementing a just transition program; the structure and organization of the industry is not conducive to long-term labour market planning. Self-employment, combined with a decentralized, fragmented industry and weak unions, makes it difficult for the United Kingdom to carry out the training needed. And UCATT has had to focus too much attention on basic survival in an industry in which its role is already marginal and its very existence uncertain. Consequently, its capacity to take a leadership role in promoting climate change initiatives – and its success in doing so – has been very limited.

CONCLUSION

The preceding discussion has examined two different national approaches to implementing climate change policies in the EU's construction sector, focusing on the role of labour. It has revealed significant differences between the two countries – differences that impact their respective abilities to implement effective climate programs in the built environment.

The German construction industry has been at the centre of the nation's policy response to climate change. Supported by a high degree of public consensus that global warming is a major issue, governments, employers, and unions have co-operated in implementing very ambitious programs to reduce energy consumption in new buildings and retrofit much of the existing building stock. The industry's capacity to implement low-carbon construction is based on training and supporting a highly skilled workforce. Germany has a high proportion of master crafts workers and a correspondingly small proportion of unskilled workers. Standards are maintained by certification regulations, which limit many areas of construction to crafts workers.

Construction employers are well organized at the national and state levels. Training is funded through a sector-wide employer levy, which covers much of the cost of wage replacement for trainees as well as other education expenses. Employers commit to providing trainees with the on-the-job experience they need to complete their apprenticeships. Qualified crafts workers have a high occupational status and expect a lifelong career in the industry. Consequently, Germany's construction workforce has the capacity to deliver innovative climate change programs.

Germany uses a social partnership model, in which the unions have a voice in a wide range of construction sector issues, including training as well as the establishment and maintenance of the standards of the various trades. While only two-fifths of construction workers are unionized, collective agreement coverage, enhanced by labour ministry directives, means that unions regulate the terms and conditions of employment of over two-thirds of workers. Unions also exercise indirect influence through members' representation in works councils. Thus, they are able to participate in shaping climate change programs both at the level of government policy and at the level of individual firms.

In contrast, the construction industry in the United Kingdom operates within a liberal-market economic framework characterized by a very low level of unionization. Its approach to industry productivity is to keep labour costs low and minimize employers' commitment to the workforce through extensive subcontracting and self-employment. Work is organized using a top-down, management approach, by which developing a highly skilled crafts workforce is not a priority. As a consequence, the training system is not well suited to providing the workforce with the new skills required to implement the greening of the UK construction sector.

UK unions play only a marginal role in influencing occupational standards and training programs, and successive governments have ignored their potential to promote climate change initiatives. Hence, the trade union movement has little ability to exercise significant influence on climate policies in construction or to oversee the training and development of the workforce in the new skills required to implement low-carbon objectives.

The conclusion is that the contribution of building workers and their unions to climate change depends on a number of factors. Broad public support for climate initiatives is important. Sound government policies can encourage industry players to give priority to these initiatives. And a labour relations system in which workers and their unions play a significant role provides opportunities for labour to contribute both to policy and to implementing low-carbon construction at the work site. Where workers and their unions play a significant role in the organization of work and the training of the workforce, as in Germany, they can – and do – positively influence the response of their industries to climate change. Conversely, where their role is marginal, as in the United Kingdom, their capacity to contribute to climate change programs is marginal. Including labour in policy development and program delivery is only one factor influencing the approach of these countries to climate change. Nevertheless, unions can make a positive contribution, but only if they have the resources and influence to make a difference.

NOTES

1. The author would like to acknowledge the contribution of Kaylin Woods, who provided background research for this chapter.
2. The figures are taken from the website of the former Federal Ministry of Economics and Technology. The ministry was renamed in 2013, and the website is no longer available.

REFERENCES

Addison, J.T., C. Schnabel, and J. Wagner. 2006. "The (Parlous) State of German Unions." Discussion Paper No. 2000. Bonn: IZA (Forschungsinstitut zur Zukunft der Arbeit/Institute for the Study of Labor). http://papers.ssrn.com/sol3/papers.cfm?abstract_id=890275.

Andrews, P., and M. Karaisl, eds. 2010. *Policy Monitor – Baseline Report 2010*. Climatico Analysis, June. http://www.climaticoanalysis.org/wp-content/plugins/download-monitor/download.php?id=3.

Auer, J., E. Heymann, and T. Just. 2008. *Building a Cleaner Planet: The Construction Industry Will Benefit from Climate Change*. Frankfurt am Main: Deutsche Bank Research.

Bartholmai, B., and M. Gornig. 2007. "Construction Industry in Germany: Production Structure and Employment – Results for 2006." Berlin: German Institute for Economic Research.

Behrens, M. 2009. "Still Married after All These Years? Union Organizing and the Role of Works Councils in German Industrial Relations." *Industrial and Labour Relations Review* 62 (3):275–93.

Billett, S., and N. Bowerman, eds. 2009. *Assessing National Climate Policy: November 2008–February 2009*. Climatico. www.climaticoanalysis.org/press-releases/Climatico_National_Assessment_Report_March2009.pdf.

Bosch, G., and J. Charest. 2008. "Vocational Training and the Labour Market in Liberal and Coordinated Economies." *Industrial Relations Journal* 39 (5):428–47.

Bosch, G., and P. Philips, eds. 2003. *Building Chaos: An International Comparison of Deregulation in the Construction Industry*. London: Routledge.

Bosch, G., and C. Weinkopf. 2010. "EC Project – Minimum Wage Systems and Changing Industrial Relations in Europe: National Report Germany." Revised version of paper prepared for EWERC, University of Manchester, 2009. Duisburg and Essen: IAQ University Duisburg-Essen.

Bosch, G., and K. Zühlke-Robinet. 2003. "The Labour Market in the German Construction Industry." In *Building Chaos: An International Comparison of Deregulation in the Construction Industry*, ed. G. Bosch and P. Philips, 48–72. London: Routledge.

Braverman, H. 1974. *Labour and Monopoly Capital*. New York: Monthly Review Press.

Brockmann, M., L. Clarke, and C. Winch. 2008. "Knowledge, Skills, Competence: European Divergences in Vocational Education and Training (VET) – The English, German, and Dutch Cases." *Oxford Review of Education* 34 (5):547–67. http://www.tandfonline.com/toc/core20/34/5#.VC3AN-evyAI.

—. 2010a. *Bricklaying Is More Than Flemish Bond: Bricklaying Qualifications in Europe*. Brussels: European Institute for Construction Labour Research.

—. 2010b. "Bricklaying Qualifications, Work and VET in Europe." *CLR News* 1:7–42. www.clr-news.org.

CAN (Climate Action) Europe and EFBWW (European Federation of Building and Woodworkers). 2011. "EFBWW and Climate Activists Join Forces to Call for Enhanced Climate and Energy Savings Action in Europe." Press release, June 20. http://www.efbww.org/default.asp?index=819&Language=EN.

Chamberlin, G., ed. 2011. *Economic & Labour Market Review* 5 (5). London: ONS. http://www.ons.gov.uk/ons/rel/elmr/economic-and-labour-market-review/may-2011online-edition/elmr.pdf.

Clough, B. 2009. "Vocational Education and Training in the UK: Social Dialogue or State Monologue?" *CLR News* 1:10–15. www.clr-news.org.

Council of the European Union. 2010. "Council Conclusions: Employment Policies for a Competitive, Low-Carbon, Resource-Efficient and Green Economy." 2053rd Employment, Social Policy, Health and Consumer Affairs Council Meeting, Brussels, 8 December.

—. 2011. *Joint Employment Report*. Brussels: Council of the European Union.

Danish Technological Institute. 2009. *Future Qualification and Skills Needs in the Construction Sector*. Policy and Business Analysis. Taastrup: Danish Technological Institute. http://ec.europa.eu/enterprise/sectors/construction/files/qualification-and-skills/final-report-july-2009_en.pdf.

Decaillon, J. 2009. "A European Approach to Tackling Climate Change." In *Working on Change: The Trade Union Movement and Climate Change*, ed. F. Scott, 47–53. London: Green Alliance. http://www.green-alliance.org.uk/resources/Working%20on%20change.pdf.

Dribbusch, H., and P. Birke. 2012. *Trade Unions in Germany: Organisation, Environment, Challenges*. Berlin: Friedrich-Ebert-Stiftung. http://library.fes.de/pdf-files/id-moe/09113-20120828.pdf.

Dromey, J., and S. Hunt. 2009. "Ensuring a Just Transition." In *Working on Change: The Trade Union Movement and Climate Change*, ed. F. Scott, 5–12. London: Green Alliance. http://www.green-alliance.org.uk/resources/Working%20on%20change.pdf.

Dupressoir, S. et al. 2007. *Climate Change and Employment: Impact on Employment in the European Union-25 of Climate Change and CO_2 Emission Reduction Measures by 2030*. Brussels: European Trade Union Confederation. https://docs.google.com/file/d/0B9RTV08-rjErMGstNWRPbkJWaWM/edit?pli=1.

EC (European Commission). 2009. "The EU Climate and Energy Package." European Commission: Climate Action. http://ec.europa.eu/clima/policies/package/index_en.htm.

—. 2011. "Roadmap for Moving to a Competitive Low Carbon Economy in 2050." http://ec.europa.eu/clima/policies/roadmap/index_en.htm.

—. 2014. "Communication from the Commission to the European Parliament, the Council, the European Economic and Social Committee and the Committee of the Regions: A Policy Framework for Climate and Energy in the Period from 2020 to 2030." Brussels: EC.

ECORYS. 2008. *Environment and Labour Force Skills: Overview of the Links between the Skills Profile of the Labour Force and Environmental Factors*. Final Report. Rotterdam: European Commission DG Environment.

EFBWW (European Federation of Building and Woodworkers). 2013. "About the EFBWW." http://www.efbww.org/default.asp?Issue=efbww&Language=EN.

—. 2014. "Building & Construction." http://www.efbww.org/default.asp?Issue=CONSTR&Language=EN.

EFBWW and BWI (Building Workers' International). 2008. "For Secure Jobs in the Building and Wood Industries: EFBWW/BWI Platform of Action for a Social and Green Europe." Brussels: EFBWW / BWI. http://www.efbww.org/pdfs/EN.pdf.

EFBWW and FIEC (European Construction Industry Federation). 2008. "Joint Work Program of the European Social Partners of the Construction Industry for the European Social Dialogue of the Construction Industry 2008–2011." Brussels: EFBWW / FIEC.

—. 2012. "Multiannual Action Programme for the Sectoral European Social Dialogue of the Construction Industry, 2012–2015." Brussels: EFBWW and FIEC.

Egan, J. 1998. "Rethinking Construction: The Report of the Construction Task Force to the Deputy Prime Minister, John Prescott, on the Scope for Improving the Quality and Efficiency of UK Construction." London: Department of Trade and Industry.

—. 2003. *Accelerating Change: A Report by the Strategic Forum for Construction.* London: Rethinking Construction. http://www.strategicforum.org.uk/pdf/report_sept02.pdf.

EMCO (Employment Committee of the European Commission). 2010a. *The Employment Dimension of Tackling Climate Change: Overview of the* State-of-Play *in Member States*. EMCO Reports Issue 3. Report endorsed by EMCO 9 October. Brussels: Employment, Social Affairs and Inclusion Directorate.

—. 2010b. *Towards a Greener Labour Market: The Employment Dimension of Tackling Environmental Challenges*. EMCO Reports Issue 4. Final report endorsed by EMCO 10 November. Brussels: Employment, Social Affairs and Inclusion Directorate.

ETUC (European Trade Union Confederation). 2005a. "Climate Change and Employment: Case of the United Kingdom." Brussels: ETUC. http://www.etuc.org/a/3676.

—. 2005b. "Review of the EU Sustainable Development Strategy." Brussels: ETUC.

—. 2006. "The New EU Sustainable Development Strategy: A Step Forward in the Right Direction." Press release, 19 June. http://www.etuc.org/a/2508.

—. 2007. "Climate Change and Employment: Case of the United Kingdom." Brussels: ETUC. http://www.etuc.org/sites/www.etuc.org/files/Rapport RU_1.pdf.

—. 2009. "Climate Disturbances, the New Industrial Policies and Ways Out of the Crisis." Brussels: ETUC. https://docs.google.com/open?id=0B9RTV08-rjErQ0hPS1poTExCeFE.

—. 2010. "Climate Change, the Industrial Policies and Ways Out of the Crisis." Brussels: ETUC. http://www.etuc.org/sites/www.etuc.org/files/Etude_EN_1.pdf.

ETUC and SDA (Social Development Agency). 2009. "The New Directive 2009/38/EC: Recommendations on Negotiating during the Transposition Period (5 June 2009 to 5 June 2011)." Brussels: ETUC / SDA. http://www.efbww.org/pdfs/EIF%20common%20recommendations%20GB.pdf.

Eurofound (European Foundation for the Improvement of Living and Working Conditions). 2011. *Industrial Relations and Sustainability: The Role of Social Partners in the Transition towards a Green Economy*. Dublin: Eurofound. http://eurofound.europa.eu/ef/sites/default/files/ef_files/pubdocs/2011/26/en/1/EF1126EN.pdf.

Gleeson, C., and L. Clarke. 2013. "The Neglected Role of Labour in Low Energy Construction: 'Thermal Literacy' and the Difference between Design Intention and Performance." Paper presented at the Work in a Warming World (W3) Conference, "Labour, Climate Change, and Social Struggle," Toronto, 29 November–1 December 2013.

Harvey, M. 2001. *Undermining Construction: The Corrosive Effects of False Self-Employment*. London: Institute of Employment Rights.

Harvey, M., and F. Behling. 2008. *The Evasion Economy: False Self-Employment in the UK Construction Industry*. London: UCATT.

Hedegaard, C. 2011. "Roadmap for Moving to a Competitive Low Carbon Economy in 2050." Presentation at Stakeholder Conference, Brussels, 17 March 2011. http://ec.europa.eu/clima/events/0032/roadmap_en.pdf.

IPCC (Intergovernmental Panel on Climate Change). 1996. *Second Assessment Report*. New York: Cambridge University Press.

—. 2013. *Climate Change 2013: The Physical Science Basis*. Cambridge: Cambridge University Press.

iXPOS. 2014. "Construction: Building on Solid Ground." http://www.ixpos.de/IXPOS/Navigation/EN/Your-business-in-germany/Business-sectors/Service-industries/construction,did=253602.html.

Jagodzinski, R. 2009. "Recast Directive on European Works Councils: Cosmetic Surgery or Substantial Progress?" *Industrial Relations Journal* 40 (6):534–45.

Kirton-Darling, J. 2011. "Roadmap 2050 – ETUC." Presentation at Stakeholder Conference, Brussels, 17 March. http://ec.europa.eu/clima/events/docs/0032/etuc_en.pdf.

Laux, E. 2008. "Paritarian Social Funds in the West European Construction Sector – A Model for Eastern Europe?" *CLR News* 1:6–24.

Laux, E., D. Schäfer, and A. Harnack. 2010. "On the Actual Financial and Economic Crisis: Its Impact on the Construction Industry and Labour Market in Germany." *CLR News* 3:34–40.

MacKenzie, R., C. Forde, A.M. Robinson, H. Cook, B. Eriksson, P. Larsson, and A. Bergman. 2010. "Contingent Work in the UK and Sweden: Evidence from the Construction Industry." *Industrial Relations Journal* 41 (6):603–21.

Morgera, E., K. Kulovesi, and M. Muñoz. 2010. "The EU's Climate and Energy Package: Environmental Integration and International Dimensions." Europa Working Papers no. 2010/07. Edinburgh: School of Law, University of Edinburgh. http://www.research.ed.ac.uk/portal/en/publications/the-eus-climate-and-energy-package%28235a1c29-3cc8-4401-8147-80af7a8e6fd1%29.html.

OECD (Organisation for Co-operation and Development). 2014. *Labour Force Statistics 2013*. Paris: OECD. http://www.oecd-ilibrary.org/employment/oecd-labour-force-statistics-2013_oecd_lfs-2013-en.

Parkes, S. 2010. "Low Carbon and Sustainability." Asset Skills Intelligence Paper 5. http://www.assetskills.org/Research/LabourMarketInformation/IntelligencePapers.aspx? - IntelliIntelligencePap5.

Recio, A. 2007. "The Construction Sector: What Model of Regulation Is It Moving Towards?" Paper prepared for the 28th International Working Party on Labour Market Segmentation, Aix-en-Provence, 5–7 July.

Richter, A. 1998. "Qualifications in the German Construction Industry: Socks, Flows and Comparisons with the British Construction Sector." *Construction Management and Economics* 16 (5):581–92.

Schneider, W. 2009. "Alliance for Work and Environment." Presentation to the Confederation of German Trade Unions (DGB), ILO/MOHRSS China Green Jobs Experience Sharing Meeting, Beijing, 30–31 May.

—. 2010. "Green Jobs Creation in Germany: By Energy Saving and Energy Efficiency in the Redevelopment of Existing Buildings and by Extension of Renewable Energy." Presentation at Cornell University, Ithaca, NY, 12 May.

Stern, N. 2006. *Stern Review: The Economics of Climate Change*. Executive Summary. Pre-publication edition. London: HM Treasury.

Toner, P. 2008. "Survival and Decline of the Apprenticeship System in the Australian and UK Construction Industries." *British Journal of Industrial Relations* 46 (3):431–38.

TUC (Trade Union Congress). 2008. "A Green and Fair Future: For a Just Transition to a Low Carbon Economy." Touchstone Pamphlet 3. London: TUC. http://www.tuc.org.uk/economic-issues/touchstone-pamphlets/social-issues/environment/green-and-fair-future-just-transition.

—. 2014. *TUC Directory 2014*. London: TUC. http://www.tuc.org.uk/about-tuc/tuc-directory-2014.

UCATT (Union of Construction, Allied Trades and Technicians) and ConstructionSkills Council. 2006. "Bilateral Agreement with UCATT." London: UCATT.

UK (United Kingdom). 2008. *Climate Change Act 2008*. c. 27. http://www.decc.gov.uk/en/content/cms/legislation/cc_act_08/cc_act_08.aspx.

—. 2009. *The UK Low Carbon Transition Plan: National Strategy for Climate and Energy*. Norwich: Stationery Office. 20 July. http://webarchive.nationalarchives.gov.uk/20100509134746/http:/www.decc.gov.uk/en/content/cms/publications/lc_trans_plan/lc_trans_plan.aspx.

UK. BIS (Department for Business, Innovation and Skills). 2010a. *Low Carbon Construction: Innovation and Growth Team*. Final Report. London: BIS. http://webarchive.nationalarchives.gov.uk/20121212135622/http://www.bis.gov.uk/policies/business-sectors/construction/low-carbon-construction-igt/innovation-and-growth-team.

—. 2010b. "TUC GreenWorkplaces – Greening the Work Environment." London: BIS. http://www.bis.gov.uk/assets/biscore/employment-matters/docs/t/10-1024-umf-2-tuc-green-workplace.

—. 2010c. "Union of Construction, Allied Trades and Technicians (UCATT) – Embedding Modernisation across the Union." London: BIS. http://bis.ecgroup.net/Publications/EmploymentMatters/UnionModernisationFund.aspx.

—. 2013. *UK Construction: An Economic Analysis of the Sector*. London: BIS. https://www.gov.uk/government/uploads/system/uploads/attachment_data/file/210060/bis-13-958-uk-construction-an-economic-analysis-of-sector.pdf.

UK. CCC (Committee on Climate Change). 2010. *The Fourth Carbon Budget: Reducing Emissions through the 2020s*. London: CCC. http://www.theccc.org.uk/publication/the-fourth-carbon-budget-reducing-emissions-through-the-2020s-2/.

—. 2013a. *Fourth Carbon Budget Review – Part 1: Assessment of Climate Risk and the International Response*. November. London: CCC. http://www.theccc.org.uk/wp-content/uploads/2013/11/1784-CCC_SI-Report_Book_single_1a.pdf.

—. 2013b. *Fourth Carbon Budget Review – Part 2: The Cost-Effective Path to the 2050 Target*. December. London: CCC. http://www.theccc.org.uk/wp-content/uploads/2013/12/1785a-CCC_AdviceRep_Singles_1.pdf.

UK. Committee of Public Accounts. 2007. *Building for the Future: Sustainable Construction and Refurbishment on the Government Estate*. Third Report of Session 2007–08. HC 174. London: Stationery Office.

UK. Department of Energy and Climate Change. 2009. *Climate Change Act 2008: Impact Assessment*. London: Department of Energy and Climate Change. http://www.climatedatabase.eu/sites/default/files/eia_climatechangeact.pdf.

UNEP (United Nations Environment Programme). 2006. "Report of the Trade Union Assembly on Labour and the Environment on the Work of Its First Meeting." Trade Union Assembly on Labour and the Environment, Nairobi, 15–17 January.

—. 2009. *Green Jobs: Towards Decent Work in a Sustainable, Low-Carbon World*. Nairobi: UNEP. http://www.unep.org/PDF/UNEPGreenJobs_report08.pdf.

Watanabe, R., and L. Mez. 2004. "The Development of Climate Change Policy in Germany." *International Review for Environmental Strategies* 5 (1):109–26.

Watson, J. 2009. "The Skills and Training Knowledge Base of VET in Britain: What the Social Partners Need to Know." *CLR News* 1:21–25. www.clr-news.org.

Weidner, H., and L. Mez. 2008. "German Climate Change Policy: A Success Story with Some Flaws." *Journal of Environment and Development* 17 (4):356–78.

CHAPTER 3

GENDERED EMISSIONS: COUNTING GREENHOUSE GAS EMISSIONS BY GENDER AND WHY IT MATTERS

Marjorie Griffin Cohen[1]

INTRODUCTION

Both climate disasters and incremental climate changes have an enormous impact on the way people live and work, and these physical changes are not confined to the most marginal areas of the world's surface, but impinge upon the developed world as well. It is also fairly clear that public policy developed to deal with climate change in too many places is woefully inadequate to deal with the magnitude of the problem (Victor 2011).[2] Such public policy has gendered implications both for its effectiveness in mitigating greenhouse gas (GHG) emissions and for different aspects of work, play, and consumption.

Feminists have not ignored the gendered effects of climate change, but, unsurprisingly, gender issues have a fairly low profile in policy discussions. At the international level, there is discussion of procedural justice as more recognition is given to the need for a variety of interests to be part of the discussion and decision-making process (IPCC 2007; Agarwal 2001). But any formal way of facilitating procedural justice has been absent from most concrete policy-making (Klinsky and Dowlatabadi 2009; Dankelman 2002). Even popular sector movements that recognize the need for inclusivity in the discussion of alternatives tend to ignore women and the distinctions by gender (Riddell 2011; Stahl, Rees, and Byers 2011).

Work in a Warming World, edited by Carla Lipsig-Mummé and Stephen McBride. Kingston: School of Policy Studies, Queen's University.

Some explain this void as being a result of two things: one relates to the need to focus on universal issues, given limited resources; and the other stresses the prominence of technological and scientific solutions in policy discussions, as opposed to the soft policies that look at social differences, particularly as they relate to incomes and opportunities (Lambrou and Piana 2006). These are the kinds of factors that frequently inhibit a gender analysis of any social issues, but climate change has a specific disadvantage that contributes to the gender blindness that occurs in research and policy, particularly in developed nations. This is the lack of visibility of gendered environmental differences or injustices and a lack of imagination about how a gender analysis could be applied in research. Gender issues are more visible in developing countries or among Aboriginal societies, and the literature about the effects of climate change and gender related to these areas constitutes the bulk of the information available on gender and climate change.[3] Because of the less-developed nature of the economies, and the close proximity of women to agriculture and the resource sectors, the effect of climate change on women's work in these countries is visible and dramatic (Brownhill 2007; Agarwal 2001; Beaumier and Ford 2010; Nelson and Stathers 2009).

Others explain the prolonged invisibility of gender issues in the climate-justice literature and lack of action in developed countries as a result of the dominance of males in the environmental movement's senior posts and the general gender blindness of the movement (Buckingham and Kulcur 2009). The actions of the state on environmental issues are conditioned by the ways they are contested by activist organizations and the kinds of issues they highlight. The fact that these organizations are usually male-dominant means that gendered issues are not explored in either identifying environmental justice issues or seeking policy solutions.

As most groups dealing with inequality know, identifying the distinctions of experience itself is crucial for being included in policy discussions. But when the differential experiences are less visible by being diffuse as a community or income group, unpacking the implications is not straightforward. While academics are increasingly interested in environmental injustices, particularly as they relate to public policy, they tend to focus on inequalities in income and/or race and neglect gender (Buckingham and Kulcur 2009). The most common understanding of environmental justice as it applies to individuals is exemplified by the definition used by the United States Environmental Protection Agency. "Environmental Justice is the fair treatment and meaningful involvement of all people regardless of race, color, national origin, or income with respect to the development, implementation, and enforcement of environmental laws, regulations, and policies" (US EPA 2014.). For the most part, gender (and age and disability) issues have been subsumed under class or income issues because it is assumed that women's issues will be covered under the treatment of low-income and poverty groups.

Many of the methods that have been used to talk about gender and climate change tend to conflate the material related to developing nations with conditions associated with developed nations. While some studies are careful about how they use examples in specific references, there is an assumption (based primarily on experiences in developing regions) that, in general, women are more vulnerable to climate change than men, have fewer resources to deal with it, and have greater work burdens as a result (Haigh and Vallely 2010; Johnsson-Latham 2007; Bjönberg and Hansson 2013).

The primary method that is used to span the national divides is the assumption that since women everywhere are, on average, poorer than men, they will, as a result, be disproportionately disadvantaged by climate change. While it is true that women in developed nations have lower incomes on average than men, gender distinctions are not confined to income distribution alone but also relate to the entire gendered experiences of our societies. One particular gap in knowledge pertains to the relationship between climate change and gender in developed nations.

The intention in this chapter is to try to understand how men and women contribute to climate change through both work and consumption. The first part will examine the gendered nature of paid work–related contributions to climate change by quantifying the emissions associated with major GHG emitters in Canada. It will also discuss the gendered nature of consumption as it relates to work, with a particular focus on the problems of unpacking gendered work and consumption in the household. (One area of work and consumption that is easily quantifiable by gender, for example, is consumption related to vehicle transportation, one of the major sources of GHG emissions in Canada.) The findings will then be analyzed in light of the significance of two areas. One will be how gendered distinctions in GHG emissions through work and consumption can be understood in relation to climate justice. The other will be the significance of gender in expanding our ideas about both green jobs and a green economy.

THE EFFECTS OF WORK ON CLIMATE CHANGE

Almost nothing has been written about the impact of work in developed nations on climate change with specific reference to the gendered nature of paid work. Most of the interest in work has been on creating green jobs, as a positive way of ensuring that government programs to reduce GHG emissions do not result in actions that increase unemployment rates.[4] The assumption is that by having a correctly designed policy, jobs can be created that have a low impact on the environment and are well paid (Lee and Carlaw 2010). This argument has been advanced to counteract the slow pace by which many developed nations are enacting effective legislation to reduce GHG emissions. North American governments' rationale for inaction, either explicitly stated or at least implied, is that unless all countries in the world adopt similar climate change policies, those countries that do so will be punished with poorer economic performance and rising unemployment rates as corporations move their production to those jurisdictions that have no or few restrictions on emissions. This is the argument that Canada uses, for example, for not highly regulating the emissions from the tar sands in Alberta (Clarke 2008) and why the US government did not sign the Kyoto Protocol (Biermann and Brohm 2005).

In broad terms, some jobs are dirty jobs that add a great deal to GHG emissions, while others have a more benign effect on the environment. This can be calculated in a variety of ways, using different types of measures of damage. While studies that look at these measures tend to talk in very broad terms, and with no reference to who does what type of jobs, it is possible to glean an idea of where gender distinctions occur.

While there are methodological (and justice) problems associated with attributing GHG emissions to specific groups of labour, counting something seems to be the primary way that gender issues are noticed. In virtually anything related to distributional impacts (such as wage inequality, occupational distribution, and poverty levels), calculating gender differences is the major way to have an issue recognized as significant. In the case of labour and GHG emission policies, bringing the gendered dimension into the discussion necessitates showing both the unequal contributions by gender and the gendered nature of mitigation or adaptation policies.

The method I have used to calculate gendered GHG emissions through work is to use the known major contributors to GHG emissions through industrial output by sector. I first examine the major sectors of the economy in Canada that contribute to GHG emissions and, through a simple weighting of the gendered composition of the labour force in that sector, calculate the gendered contribution in each sector.

PAID EMPLOYMENT – GHG EMISSIONS BY INDUSTRIAL SECTOR

The energy sector is the largest source of GHG emissions, accounting for 81 percent of total emissions in Canada; this includes all energy and heat generation and consumption in households, businesses, and transportation (Environment Canada 2011). The remaining emissions come primarily from the agricultural sector (8 percent) and the industrial processes sector (8 percent).

Energy production itself (from electricity as well as from oil and gas) accounts for 37 percent of total GHG emissions. While electricity production in Canada is largely hydro-based and accounts for only 16 percent of total GHG emissions, this country is the source of a particularly dirty form of oil, which is derived from the tar sands in Alberta. Production in the oil and gas sector alone accounts for 21 percent of total GHG emissions. Table 3.1 below shows the gender breakdown of those employed in the energy sector in 2007 compared to the gender breakdown in the labour force as a whole.

TABLE 3.1
Labour Characteristics of the Energy Sector in Canada, 2007 (% of total employed)

	Canadian labour force (%)	*Oil and gas (%)*	*Electricity (%)*
Men	52	72	75
Women	48	28	25

Source: Author's compilation using data from Statistics Canada (2007).

In the sections that follow, I calculate the gendered distinctions in the GHG emissions produced in three categories: industrial production, transportation, and residential. As we will see, industrial production accounts for 66 percent (455,000 megatonnes, or MT) of the total, transportation 28 percent (195,000 MT), and residential 6 percent (41,000 MT). These calculations are drawn from the National Inventory Report 1990–2011 (Environment Canada 2013).

Industrial Production

Table 3.2 below shows the major sources of GHG emissions by industry in Canada in 2010. (Transportation, the largest single sector contributing to GHG emissions, is omitted from this table and will be dealt with in the next section because the method used for calculating the gender share in transportation is slightly different.)[5]

As Table 3.2 below shows, industrial production in Canada is highly gendered, with women dominant in industries that have lower GHG emissions than those dominated by men. So, for example, commercial and institutional industries account for only 4 percent of total GHG emissions and are sectors dominated by women. The industries with the highest GHG emissions are the energy sector (see the gender breakdown presented in Table 3.1) and agriculture, where women comprise 30 percent of the labour force. In manufacturing and industrial processes, men again dominate in most of the heavy emitters (iron, steel, and metal production), cement and mineral products, and other subsectors. The only subsector with high emissions and a high proportion of women's employment is the chemical subsector, where women account for 41 percent of the labour force.

Altogether, the gendered breakdown of GHG emissions in manufacturing and industrial processes is highly skewed, with the male share accounting for 76 percent and the female share 24 percent.

TABLE 3.2
Gender Share of GHG Emissions by Industry in Canada, 2010 (MT)

	Total		*Labour*		*MT*	
GHG category	*MT*	*%*	*Female (%)*	*Male (%)*	*Female*	*Male*
Overall GHG emissions	692,000					
Major industrials						
Electricity and Heat	101,000	15	24	76	24,240	76,760
Fossil Fuel Production	53,000	8	10	90	5,300	47,700
Mining, Oil, and Gas*	96,800	14	18	82	17,424	79,376
Manufacturing Industries/ Industrial Processes**	93,100	13	28	72	26,068	67,032
Iron and Steel	22,890	3	11	88	2,518	20,143
Chemical	16,500	2	41	59	6,765	9,735
Pulp and Paper	6,460	1	23	77	1,486	4,974
Cement/Lime/Minerals	12,070	2	92	8	966	11,104
Other Manufacturing	35,700	5	28	72	9,996	25,704
Construction	1,490	0	11	89	164	1,326
Commercial & Institutional	28,400	4	56	44	15,904	12,486
Agriculture	59,260	8	30	70	16,800	39,200
Waste	22,000	3	19	81	1,480	17,820
Total, major industrial emitters	455,050	66	24	76	111,058	343,992

*Includes fugitive sources.
**Includes Manufacturing Industries.
Sources: Environment Canada (2012); Statistics Canada (2012a, 2012b, custom data).

GENDERED CONSUMPTION

Transportation

When issues of consumption are examined, the main focus of gender differences tends to be on transportation, primarily because this form of consumption looms so large in GHG emissions. In Canada, transportation accounts for about 28 percent of total GHG emissions, the single largest category of energy consumed, and within this sector, road transportation accounts for 69 percent of total emissions for all transportation (or 19 percent of the total).[6]

In the United States, men are more likely than women to drive long distances to work:[7] about 3.5 million people have a travel time of four hours per day, and two-thirds of these are men. In Sweden, it is estimated that men account for at least about 75 percent of all driving. About 6.9 million cars are registered in the country, of which women own only 1.7 million, or 25 percent, and women represent about two-thirds of all households where no one has a driver's licence (Johnsson-Latham 2007, 53).

My calculations for Canada show clearly that men dominate in the number of kilometres driven and account for most of the vehicle emissions. As can be seen from Table 3.3 below, women accounted for about 34 percent of kilometres driven in 2009, and almost all of these occurred in passenger vehicles. Passenger vehicles emit fewer GHGs than trucks, especially very big trucks, which are almost exclusively driven by men.

TABLE 3.3
Kilometres Driven by Vehicle Type and Gender in Canada, 2009 (millions)

Type of vehicle	*Total*	*Males*	*Females*	*Females (%)*
Vehicles up to 4.5 tonnes	302,959	208,083	94,877	31
Trucks 4.5 to 14.9 tonnes	8,242	8,153	89	1
Trucks 15 tonnes and over	21,231	20,673	558	3
Total, all vehicles	332,432	236,908	95,524	29

Sources: Author's compilation from Environment Canada (2012, Table A12–2) and Statistics Canada (2009, Table 405-0073).

By applying women's proportion of driving by vehicle type (Table 3.3) to emissions by vehicle type (Table 3.4 below), it is possible to show that women as a group contribute about 11 percent of GHG emissions from driving vehicles. I have calculated this in the following way: GHGtf = $V_1 (f_1) + V_2 (f_2) + V_3(f_3)$.

GHGft = GHG emissions for female vehicle owners; V_1 = GHG emissions from vehicles up to 4.5 tonnes; V_2 = trucks 4.5–14.9 tonnes; V_3 = trucks over 15 tonnes; $f_{1,2,3}$ = percentage of women driving each class of vehicle.

TABLE 3.4
GHG Emissions by Vehicle Type and Gender in Canada, 2009 (MT)

	Total	*Females (km, %)*	*Males (km, %)*	*GHG (females)*	*GHG (males)*
Vehicles up to 4.5 tonnes	41,701	31	69	12,927	28,774
Trucks 4.5 to 14.9 tonnes	44,890	1	99	449	44,407
Trucks 15 tonnes and over	47,110	3	97	1,413	45,697
Total, all vehicles	133,701			14,789	118,878
Total, all vehicles (%)				11%	89%

Sources: Author's compilation from Environment Canada (2012, Table A12–2) and Statistics Canada (2009, Table 405-0073).

Figure 3.1 below shows GHG emissions by gender and vehicle type for 2009 in Canada, revealing the large gender discrepancy for all types of vehicles as well as the GHG emissions associated with them. Figure 3.2 below shows the total proportion of emissions by gender in Canada for the same year.

The gender breakdown is also available for domestic civil aviation, but not for railways, navigation, or other forms of transportation. Total GHG emissions from civil aviation are 6,200 MT. According to the Commercial Aircraft Travel Survey of Residents of Canada 2010, women account for 54 percent of air travel and men 45 percent. Through air travel, then, women are responsible for 3,370 MT and males 2,829 MT (Statistics Canada 2010).

FIGURE 3.1
GHG Emissions by Vehicle Type and Gender in Canada, 2009 (kilotonnes)

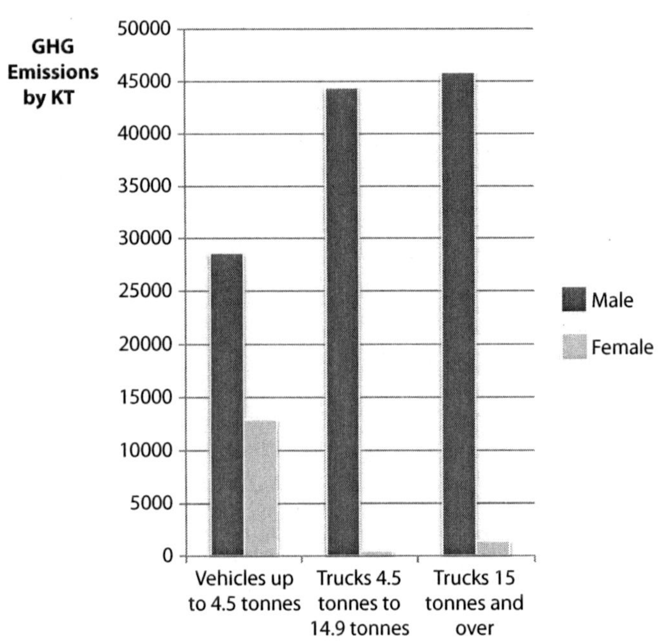

Sources: Environment Canada (2012, Table A12–2; 2011, Table S-2).

FIGURE 3.2
GHG Emissions from All Road Vehicles by Gender in Canada, 2009 (%)

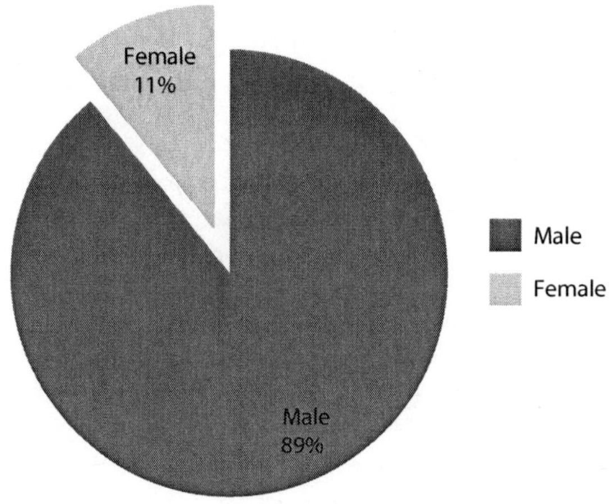

Source: Author's compilation from Environment Canada (2012, Table A12–2) and Statistics Canada (2009, Table 405-0073).

Residential

One major difficulty in discussing the gendered distinctions in the impact of work on the climate is the problem of separating the act of consumption from activities that constitute work. Consumption is integral to concepts related to work (such as gasoline consumption) and is especially relevant in places like the household. Deciding what constitutes work in the household is a conceptual problem that separates it from work that is marketized. Certain kinds of household-related consumption have gendered patterns that relate to the work roles in or for the household. In this chapter, I am primarily concerned with the gendered division of labour in the household because it has important implications for public policy responses to climate change.

In broad sweeps, the literature tends to find that virtually everywhere throughout the world, men account for greater consumption and greater carbon dioxide (CO_2) emissions than women (Johnsson-Latham 2007). This consumption pattern is primarily associated with the greater mobility enjoyed by a larger proportion of men. Analyzing gendered consumption in Sweden, for example, shows that men's consumption of transportation is double that of women. Other areas where male consumption is higher than female in Sweden are eating out, alcohol and tobacco, and leisure activities related to sports. The areas where female consumption is greater than male relate to household consumption, where women make the majority of purchases for consumer goods; clothing and shoes; health and medical care; and books, newspapers, and media (see Appendix 3.1 near the end of this chapter).

However, the Swedish study does not show the energy content of the gendered consumption baskets, so for areas other than transportation, the GHG emissions impact of consumption by gender is not clear. As a result, gaining a clear picture of gendered consumption in the household is not straightforward. In Canada, Statistics Canada, our major data collection body, makes no attempt to understand these distinctions. In the United States, there is an interest in the subject, but the data include observations of consumption only for men and women living alone; thus, it does not permit an understanding of spending by gender in households with both men and women. One attempt to research this issue is the US Consumer Expenditure Survey, which examines gender differences between single men and single women; this is the only data so far to show gender differences in household spending. The survey does not collect data on which member of households consisting of both men and women purchased an item, and the spending patterns of single men and women may be very different than households where joint decisions are made (Shipp 1987).

Class or income level is the issue that is usually found to affect the level of consumption in any country (OECD 2008). The wealthier a country is, the higher the level of consumption, and within countries, the wealthier a household is, the higher the consumption level. Measures of consumption for climate change purposes usually invoke the carbon footprint, a measure of the quantity of CO_2 emissions associated with an activity (Kitzes and Wackernagel 2009).[8] Sweden, for example, has a considerably higher level of energy consumption than the European Union (EU) average, with the average Swedish consumer driving a more energy-intensive car, living in a larger space, and eating more meat than other EU citizens (Johnsson-Latham 2007). Similarly, the average person in the United States consumes a great deal more than most people in the world, and those in Canada have among the highest per capita energy consumption.[9] In Canada, high energy use is usually explained by the cold climate and massive size of the country.

Within nations, consumption is usually correlated with income levels, which in turn can be associated with marital status and paid employment by gender (Mackenzie, Messinger, and Smith 2008). However, information on this varies from country to country. A Dutch study showed that two-income families, or families in which women had paying jobs, tended to have higher consumption levels as a result of higher incomes and the increased time pressures from having both paid and unpaid work. A two-income household used more energy than a household in which either the woman did not work or was living alone (Clancy and Roehi 2003).

A UK study showed somewhat different results. As elsewhere, energy and transport were the biggest contributors to the footprint of households.

Rural and adult households with few members had significantly higher energy use than did urban and suburban households as well as households with many members. The direct household energy use related to heating, cooking, and fuelling cars, but also considered was the indirect energy use in consumption associated with the production, processing, and distribution of food. This study also found that a higher household income affected energy use, most notably for travel (Caird and Roy 2006). In Canada, evidence shows that the ecological footprint of high-income households is substantially greater than other households and that with the exception of expenditures on food, consumption in every category increases steadily as incomes increase (Mackenzie, Messinger, and Smith 2008).

At least one study in the United States, however, finds little evidence to indicate differences between male- and female-headed households, and in countries where more equal economic conditions exist, women tend to adopt male-type lifestyles and consumption patterns (Lambrou and Piana 2006).

Gender Totals and Percentages

By aggregating the GHG emissions by gender from the three sectors discussed above – industrial production, transportation, and residential – one is able to arrive at an approximate sense of the gendered nature of GHG emissions in Canada. (See Table 3.5 below.) In the absence of a method for calculating the gendered nature of GHG emissions in a household in Canada, I have divided them equally in Table 3.5, so that women and men each account for 50 percent of residential GHG emissions, or 20,500 MT.

TABLE 3.5
GHG Emissions by Gender in Canada, 2009 (MT)

Category	*Total*	*Women*	*Men*
Transportation	455,050	111,058	343,992
Road	133,701	14,789	118,878
Air	6,200	3,370	2,829
Residential	41,000	20,500	20,500
Total	635,951	149,717	486,199
Total (%)		23.5%	76.5%

Note: Overall GHG emissions in Canada = 692,000 MT. This table includes only those categories where it was possible to ascertain gender differences. Transportation includes only road and air.

Sources: Environment Canada (2012, Table A12–2; 2011, Table S-2); Statistics Canada (2012a, Table 282-0008; 2012b; custom data).

DISCUSSION

The major question is whether the gendered contributions to climate change matter in any way. There are several ways to examine this question, but I will focus on three issues that may be significant. One relates to issues of climate justice, where it is often understood that those responsible for GHG emissions have a greater responsibility and thus greater obligations than those who do not. The second issue relates to public policy initiatives to reduce GHG emissions; knowing the gendered impact of these policies can be important for climate justice. The third issue, and one that may ultimately be the most significant, relates to climate change initiatives that are broader than those currently envisioned. As this discussion will point out, I believe that the climate justice implications are less significant than the implications of gendered distinctions in GHG emissions for the economy itself. I hope to show that expanding the notion of what constitutes a green job or a green economy can be accelerated if issues of gender are taken into account. This could be crucial for moving toward a cleaner, fairer economy altogether.

Climate Justice

Academic discussions of climate change and issues of ethics focus almost exclusively on distributive and procedural justice as it pertains to nations' climate reduction policies (e.g., Klinsky and Dowlatabadi 2009; Posner and Sunstein 2008; Adger 2006).[10] Some international groups have posited principles for climate justice that go beyond ethical issues among nations and recognize the significance of identifying certain identifiable populations as crucial for consideration in public policy issues. The Environmental Justice movement in the United States, for example, cites ten principles for climate justice policies in that country, and it specifically mentions low-income workers, people of colour, indigenous peoples, and future generations. It also calls for community participation in decisions related to climate change (Environmental Justice Movement 2002). The Bali Principles of Climate Justice are more extensive, listing 20 principles that include various types of justice focusing on the poor, women, rural, and indigenous peoples (International Climate Justice Network 2002).

One academic study that is useful for dealing with the applied ethics involved in climate policy is by Sonja Klinsky and Hadi Dowlatabadi (2009). In this review of the literature on climate change justice, five basic principles emerge as a guide to the ethics of climate change policy. These are as follows:

- Causal responsibility (the polluter pays).
- Preferential treatment based on need (related to the ability both to pay for emissions reductions and to recover from climate change impacts, with a focus on the most vulnerable).

- Equal entitlements (to protect from climate impacts).
- Equal burdens (to bear climate policy costs). The assumption is that all people have equal moral responsibility, and there is no reason why some should have a heavier burden than others.
- Procedural justice (i.e., representation by all who have a stake in the outcomes of climate policy).

Clearly, some of these principles can be at odds with others: causal responsibility is at odds with the equal burden principle, and equal entitlements can be at odds with preferential treatment. These are the types of ethical issues that are common in discussions about crafting public policy, and they have been addressed in various types of human rights legislation.

I will focus my analysis on causal responsibility and what this might mean for certain kinds of policy initiatives. The main point of causal responsibility is that "the mismatch between beneficiaries of fossil energy and victims of climate impacts gives the beneficiaries special obligations" (Klinsky and Dowlatabadi 2009, 90). Normally, this is associated with the pain experienced in less developed nations and the benefits conferred on developed nations. The climate justice message is that it is not justifiable that those who have benefited and those who have not benefited bear equal obligations. However, the idea of identifying who is responsible for climate change is problematic, particularly when people individually benefit from increased GHG emissions in different ways, such as through their work, consumption, or profits.

The major problem with assigning a gendered position to causal responsibility is that the beneficiaries, by gender, are not easily determined. While men benefit, often because of higher rates of employment and wages when working in dirty industries, our economic reality is that what is produced is diffused throughout the population and is often essential to maintaining life or a standard of living. Also, men and women live together and share material goods in the household. While this may not be equal sharing, as discussed above, until consumption patterns reveal that men consume in ways that are more detrimental to the environment than women, for most purposes the assumption of equal benefit from male employment in dirty industries makes sense.

Because the household is shared space, assigning gendered weight to consumption intensity is problematic. Thus, while men may have a larger carbon footprint because they drive more and farther, in more gas-guzzling machines, the social, gendered nature of the division of labour, rather than consumption decisions, largely conditions these work-related aspects of consumption. All consumption practices are embedded in a social context, and households are collective entities, where some consumption practices do relate to individual preferences, but most household-consumption decisions relate to collective use and are subject to the usual processes of organizing household behaviour. The structural nature of the economy greatly affects how closely households shape their consumption patterns according to income distribution, employment levels, and the gendered division of labour (Schultz and Stiess 2009). The important point to be taken from this is that gendered-consumption intensity may in some respects reflect individual choices, but for the most part is shaped by larger economic and social issues. This does not, however, mean that public policy decisions relating to household consumption will not have significant gendered impacts.

Some Public Policy Considerations

Having different information about the gendered nature of climate change issues could be very important for understanding the impact of various types of public policy designed to reduce GHG emissions. I will use as examples two of the most frequently used policy instruments that are designed to change people's behaviour in order to reduce GHG emissions. These are time-of-use electricity metering and carbon taxes on gasoline. Because of the gendered division of labour and different consumption habits, the respective policies may have different impacts.

Smart Meters and Time-of-Use Electricity Policy

Time-of-use electricity metering is gaining popularity, particularly in jurisdictions that rely on fossil fuels to generate electricity. In order to save an electricity company money that it would otherwise spend to create expensive new generation, several types of strategies are employed to reduce energy use. One important strategy that would smooth out energy use (and avoid building for peak periods) is to encourage people to reduce their household concentration of energy use at specific times (e.g., from 7 to 9 a.m. and from 5 to 10 p.m.) to other times of the day. This is usually done by providing reduced rates for off-peak periods. The impact this has on certain types of household labour can be substantial. The most frequently cited example of how this strategy can reduce a household's electricity bill is to shift activities like doing laundry to late-night hours, when the demand for electricity drops dramatically.

Setting public policy on climate change creates many new daily housework and caring responsibilities without any understanding of the impact on household work. As one analyst has noted, "With the rise of public campaigns for environmental awareness, those who manage households ... are expected to be more diligent in adopting time consuming green practices like recycling and precycling" (MacGregor 2006, 69). All of these issues become even more significant in a neoliberal climate, where appropriate public services are being reduced without new ones being created to meet the additional needs of families at a time when two incomes are the norm.

This is not to imply that a government's environmental policy should not apply to the realm of the household, but rather that there needs to be an increased understanding of the work burdens these changes imply. The method to be used here would be time-use studies of gendered work in the household. Gaining knowledge of these gendered impacts might allow different kinds of public policy initiatives to be considered. What I have in mind are policies related to paid work that affect the gendered division of labour in the home. It has long been recognized that the higher demands of male workforce participation by hours worked have contributed to a negative effect on contributions to household labour. The policy implications, then, would have not only environmental but also market-oriented and social implications. Reducing hours of work, hourly productivity, and the employment-to-population ratio is thought to be a solution to the environmental problem posed by long hours of work (Hayden and Shandra 2009). Reducing hours of work by increasing productivity would shift remuneration-to-time affluence and could have a marked impact on household workloads by gender.

So far, the examination of household behaviour seems to be of interest to climate change researchers less because of the justice issues involved than because of the ways in which different consumer behaviours can be affected by public policy. Public policy-makers and policy analysts are interested in knowing what affects people's attitudes and decisions and how households respond to environmental policies (OECD 2008). Thus, while certain kinds of policies, such as user charges for waste disposal, can be identified as being regressive, with poorer households assuming a greater proportion of the policy, little is known about the increase in household work and who does it when policy initiatives result in more work-intensive processes.

Carbon Taxes

The literature on the effects of carbon taxes tends to focus on differential impacts based on class. A carbon tax is a tax on the purchase of fuels, including gasoline, diesel, natural gas, and coal – all fuels emitting carbon. In the body of literature on this subject, households are treated as undifferentiated units, and gender distinctions are not usually considered (Metcalf 2008). Interest usually focuses on whether a disproportional tax burden treats the poor unjustly. When this is understood, there are attempts to encourage governments to design the tax in such a way as to correct its regressive nature (Lee and Carlaw 2010).

One exception to the approach in the academic literature is the work done by Nathalie Chalifour. Her study examines carbon taxes as they were instituted in two provinces in Canada: British Columbia and

Quebec. In British Columbia, the government sought popular approval by making the tax revenue neutral, meaning that it would be instituted not to generate revenue but rather to change consumer behaviour. This meant that many tax reductions and rebates were built into the design, as were the promises for using any surplus revenues. Chalifour uses the causal responsibility principle of social justice as her framework for understanding gender distinctions.

> Just as it would be inequitable to expect the same level of emissions reductions from countries that have contributed little to creating the climate change problem, it would be inequitable to design policy responses to climate change that place a greater burden on women than on men. (Chalifour 2010, 186)

She also makes the point that any analysis of environmental tax policy must include an examination of the tax as well as any complementary policies (that is, income tax deductions) and decisions pertaining to the use of the revenue generated by the tax. The "purpose of this goal is to ensure, at a minimum, that inequality between women and men is not perpetuated by the policy and, ideally, to seek out carbon tax policies that are capable of promoting gender equality" (Chalifour 2010, 191). Chalifour's conclusion is that women are disproportionately affected because they are, on average, poorer than men and, therefore, would pay a greater proportion of their income on the tax than men. This may well be the case, and it is an important contribution to the understanding of the complexities of climate change policy.

My intention is to build on this approach by urging greater research into gendered distinctions in both the contribution to climate change and the impact of public policy designed to either mitigate or adapt to climate change. For example, if we assume that a household is shared space with equal responsibilities for climate change, the greatest impact of a carbon tax on individuals by gender may be on vehicle drivers. If this is true (and it is by no means certain), the greatest impact would be on men, who drive considerably more than women. Determining who, in this case, would be most affected by gender considerations is complex, and new ways of looking at disparities besides income distribution need to be considered. Information about the incidence of the tax would also benefit from a clear understanding of the types of household (whether headed by men or women, age of each person, etc.) and energy use.

Wider Implications for Green Jobs and a Green Economy

Ultimately, knowing the gender contribution to GHG emissions lends weight to the argument that including gender considerations in public policy decisions about climate change could be a significant dynamic for a very different policy approach to the economy. Most current ideas about green jobs and a green economy focus fairly narrowly on reducing GHG emissions from the dirtiest industries. They include paying substantial subsidies to dirty industries in order to green them and creating entirely new industries in the energy sector to displace the reliance on fossil fuels. When gender issues are considered at all in this framework, it is from the point of view of how to move women into the typically male-dominated jobs in these industries. Where training for green jobs occurs, it is usually in the building trades, transportation, or the energy sector.

The push, then, is to include women in the training initiatives that are normally the purview of men in the hope that women will eventually receive jobs through green initiatives. What tends to be ignored is the daunting task of not only training women but having them accepted in non-traditional jobs in the skilled trades (McFarland 2013). The struggles with this over the years have been the object of much feminist activity, and while there are significant initiatives from time to time, the overall tendency is for little to change (Cohen 2003). In the new green industries that are created (such as wind, solar, and biomass), the overwhelming tendency is for the distribution of labour by gender to continue along the same pattern as in the energy industry at large (Cohen and Calvert 2013).

The initiatives related to expanding the number of women in green industries continue to dominate the discussion of gender issues related to green jobs, but there is an increasing interest in expanding the concept of green jobs altogether so that a wider range of occupations can be included in policy initiatives. This is often referred to as greening jobs, and it includes ideas about how all existing jobs can be less environmentally damaging (Lipsig-Mummé 2013).

This is a considerable shift from the current method of treating employment that is not only environmentally damaging but also largely segregated by gender. In addition, serious discussions are beginning to take place, even in more traditional arenas like the United Nations, about expanding the concept of a green economy and rethinking traditional ideas about how economies are structured and oriented primarily toward economic growth. The United Nations Environment Programme has gone so far as to state in its Green Economy Initiative (UNEP n.d.) that there is disillusionment with our prevailing economic paradigm and a belief that the economic and ecological crisis can be overcome by fostering a green economy.

In many respects, it may appear utopian to envision an economy that can protect the environment but still provide the crucial materials and services that people need. Clearly, moving in this direction would be almost impossible with the vested interests of the current economic structures so firmly in place. But if governments could be convinced to promote areas where there are real needs, primarily among the caring sector, and reduce all of the government support that goes to the energy and material goods sectors, the beginnings of a shift in economic emphasis could occur. My main point in this regard is that making caring work more visible by recognizing its significance to the general economic health of a nation could stimulate a shift in the structure of an economy. Identifying sectors of the economy that are already green – that is, sectors where women are concentrated – and giving jobs in these sectors as much prominence in the discussion of what is green as those traditionally oriented toward economic growth would be a first step in rethinking economic priorities.

CONCLUSIONS

There are gendered differences in the work associated with climate change. Men, as a group, are more involved than women are with work that contributes to GHG emissions. This is evident by their direct work in industries that are identified as the major sources of GHG emissions in Canada and by their much greater contribution to GHG emissions through vehicle use. Less clear is the division of responsibility for GHG emissions from labour in the household. This chapter has argued that until significant time-use studies of gendered work in the household are undertaken, we need to assume that responsibility for this source of emissions is shared. This does not mean, however, that household labour is unaffected by public policy related to GHG emissions. Public policies frequently impose increased labour burdens on households, and, without appropriate offsets, they could have significantly unequal impacts by gender.

But the most significant implications of understanding gender contributions to GHG emissions will be the shifts that could occur in dealing with issues of climate change. While the production of clean energy and energy efficiency as well as the reduction in emissions from dirty industries and vehicle use are all clearly essential, more attention must be given to boosting those areas of the economy that are inherently less damaging to it, but provide essential human needs. Such a shift would have the most positive gender implications.

APPENDIX 3.1

Consumption Expenditure in Sweden per Capita, 2004 (krona)

Type of consumption	*Men*	*Women*	*Men's consumption as a ratio of women's*
Eating out	2,010	1,370	3:2
Alcohol	360	160	2.5:1
Tobacco	620	380	2:1
Consumer goods	190	820	1:4
Hygiene products	40	800	1:20
Household services	620	1,220	1:2
Clothing and shoes	3,010	4,720	2:3
Health and medical care	1,470	2,450	2:3
Transportation	1,350	740	3:2
Car repairs and maintenance	670	380	2:1
Leisure activities	2,800	2,650	1:1
Sports	1,350	970	3:2
Books, newspapers, TV licence	430	690	2:3

Source: Adapted from Johnsson-Latham (2007, 39).

NOTES

1. This study is part of the Work in a Warming World (W3) research project. The author appreciates the financial support received for this project through a Social Sciences and Humanities Research Council of Canada Community-University Research Alliances grant. A version of this chapter appeared earlier as "Gendered Emissions: Counting Greenhouse Gas Emissions by Gender and Why It Matters" (Cohen 2014). The author would like to thank the following for their research assistance on this project: Heather Whiteside, Matt Pelling, and Greg O'Brien.
2. Canada reduced its GHG emissions between 2007 and 2010. Much of this can be attributed to the decreased production associated with the economic recession, although the shift away from using coal for electricity production in some provinces also contributed to the improvement. However, the long-term trajectory shows that the pattern of reduction is not expected to continue (Environment Canada 2011, 2013).
3. Whiteside and Cohen (2012) have produced a bibliography related to gender and climate change.
4. For example, Blue Green Canada, the alliance among environmentalists, civil society, and trade unions, describes its work this way: to "advocate for working people and the environment by promoting solutions to environmental issues that have positive employment and economic impacts" (http://www.bluegreencanada.ca/about).
5. The calculations of the gendered (i.e., female) share of GHG emissions from major industrial emitters (GHG female major industrials = GHGfmi) are as follows: GHGfmi = $I_{1\text{-}12}(F_{1\text{-}12})$. I_1 = GHG emissions from electricity and heat generation; I_2 = fossil fuel production and refining; I_3 = mining and oil and gas extraction; $I_{4\text{-}8}$ = manufacturing heavy emitters (iron, steel, and non-metals; chemical; paper and pulp; cement; other manufacturing); I_9 = Construction; I_{10} = Commercial & Institutional; I_{11} = Agriculture; I_{12} = waste; $F_{1\text{-}12}$ = percentage of women working in these industrial sectors.
6. Calculated from Environment Canada (2011, Table S-1).
7. It should be noted that information available is not consistent across countries. In Canada, information about vehicle use is available by vehicle type and kilometres driven, but not by the division between work and other activities.
8. The carbon footprint is a component of the ecological footprint. The ecological footprint is a concept used in resource accounting to measure how much productive land and sea is used by an individual, nation, or type of activity and is related to the total available (Rees 2006).
9. The United States and Canada are among the highest energy users in the world. According to World Bank statistics, Canada used 7,243 kilograms of oil equivalents per person in 2012, and the United States used 6,793 (http://data.worldbank.org/indicator/EG.USE.PCAP.KG.OE).
10. *Distributive justice* refers to the ways that both the burden of climate change and its solutions are ethically part of public policy. *Procedural justice* refers to the inclusion of all who are affected by climate change in designing public policy to deal with it.

REFERENCES

Adger, W.N., ed. 2006. *Fairness in Adaptation to Climate Change.* Boston: MIT Press.

Agarwal, B. 2001. "Participatory Exclusions, Community Forestry, and Gender: An Analysis for South Asia and a Conceptual Framework." *World Development* 29 (10):1623–48.

Beaumier, M.C., and J.D. Ford. 2010. "Food Insecurity among Inuit Women Exacerbated by Socioeconomic Stresses and Climate Change." *Canadian Journal of Public Health* 101 (3):196–201.

Biermann, F., and R. Brohm. 2005. "Implementing the Kyoto Protocol without the USA: The Strategic Role of Energy Tax Adjustments at the Border." *Climate Policy* 4:289–302.

Bjönberg, K.E., and S.O. Hansson. 2013. "Gendering Local Climate Adaptation." *Local Environment* 18 (2):217–32. http://dx.doi.org/10.1080/13549839.2012.729571.

Brownhill, L.S. 2007. "Gendered Struggles for the Commons: Food Sovereignty, Tree-Planting, and Climate Change. *Women & Environments International Magazine* 4:34–37.

Buckingham, S., and R. Kulcur. 2009. "Gendered Geographies of Environmental Justice." *Antipode* 41 (4):659–83.

Caird, S., and R. Roy. 2006. "Household Ecological Footprints – Demographics and Sustainability." *Journal of Environmental Assessment Policy and Management* 8 (4):407–29.

Chalifour, N. 2010. "A Feminist Perspective on Carbon Taxes." *Canadian Journal of Women and the Law* 21 (2):169–212.

Clancy, J., and U. Roehi. 2003. "Gender and Energy: Is There a Northern Perspective?" *Energy for Sustainable Development* 7 (3): 44–49.

Clarke, T. 2008. *Tar Sands Showdown: Canada and the New Politics of Oil in an Age of Climate Change.* Toronto: James Lorimer.

Cohen, M.G., ed. 2003. *Training the Excluded for Work: Access and Equity for Women, Immigrants, First Nations, Youth and People with Low Income.* Vancouver: University of British Columbia Press.

—. 2014. "Gendered Emissions: Counting Greenhouse Gas Emissions by Gender and Why It Matters." *Alternate Routes* (25):55–80.

Cohen, M.G., and J. Calvert. 2013. "Climate Change and Labour in the Energy Sector." In *Climate@Work,* ed. C. Lipsig-Mummé, 76–104. Halifax: Fernwood Publishing.

Dankelman, I. 2002. "Climate Change: Learning from Gender Analysis and Women's Experiences of Organizing for Sustainable Development." *Gender and Development* 10 (2):21–29.

Environmental Justice Movement. 2002. "10 Principles for Just Climate Change Policies in the U.S." http://www.ejnet.org/ej/climatejustice.pdf.

Environment Canada. 2011. *National Inventory Report 1990–2009: Greenhouse Gas Sources and Sinks in Canada.* Executive Summary. Cat. No. En81-4/1-2009E-PDF. Ottawa: Environment Canada. http://www.ec.gc.ca/Publications/default.asp?lang=En&xml=a07097EF-8EE1-4FF0-9aFB-6c392078d1a9.

—. 2012. *National Inventory Report 1990–2010: Greenhouse Gas Sources and Sinks in Canada – The Canadian Government's Submission to the UN Framework Convention on Climate Change. Part 3*. Ottawa: Environment Canada. http://publications.gc.ca/collections/collection_2012/ec/En81-4-2010-3-eng.pdf.

—. 2013. *National Inventory Report 1990–2011: Greenhouse Gas Sources and Sinks in Canada – The Canadian Government's Submission to the UN Framework Convention on Climate Change*. Executive Summary. Ottawa: Environment Canada. http://ec.gc.ca/Publications/default.asp?lang=En&xml=A07ADAA2-E349-481A-860F-9E2064F34822.

Haigh, C., and B. Vallely. 2010. *Gender and the Climate Change Agenda: The Impacts of Climate Change on Women and Public Policy*. London: Women's Environmental Network.

Hayden, A., and J.M. Shandra. 2009. "Hours of Work and the Ecological Footprint of Nations: An Exploratory Analysis." *Local Environment* 14 (6):575–600.

International Climate Justice Network. 2002. "Bali Principles of Climate Justice." http://www.indiaresource.org/issues/energycc/2003/baliprinciples.html.

IPCC (Intergovernmental Panel on Climate Change). 2007. "Summary for Policymakers." In *Climate Change 2007: Impacts, Adaptation and Vulnerability – Contribution of Working Group II to the Fourth Assessment Report of the Intergovernmental Panel on Climate Change*, ed. M.L. Parry, O.F. Canziani, J. Palutikof, J.P. van der Linden, and C.E. Hanson, 7–22. Cambridge: Cambridge University Press.

Johnsson-Latham, G. 2007. *A Study on Gender Equality as a Prerequisite for Sustainable Development: What We Know about the Extent to Which Women Globally Live in a More Sustainable Way Than Men, Leave a Smaller Ecological Footprint and Cause Less Climate Change*. Report to the Environment Advisory Council. Stockholm: Environment Advisory Council, Ministry of the Environment.

Kitzes, J., and M. Wackernagel. 2009. "Answers to Common Questions in Ecological Footprint Accounting." *Ecological Indicators* 9 (4):812–17.

Klinsky, S., and H. Dowlatabadi. 2009. "Conceptualization of Justice in Climate Policy." *Climate Policy* 9 (1):88–108.

Lambrou, Y., and G. Piana. 2006. "Gender: The Missing Component of the Response to Climate Change." Rome: Food and Agriculture Organization of the United Nations, Gender and Population Division, Sustainable Development Department. http://www.fao.org/publications/search/en/?sel=e3F9Z2VuZGVyIHRoZSBtaXNzaW5nIGNvbXBvbmVudCQkJGZkcl9jX2NvbmNlcHQ6Imh0dHA6Ly9haW1zLmZhby5vcmcvYW9zL2Fncm92b2MvY18zNDgzNSI%3D.

Lee, M., and K.L. Carlaw. 2010. *Climate Justice, Green Jobs and Sustainable Production in BC*. Vancouver: Canadian Centre for Policy Alternatives.

Lipsig-Mummé, C. 2013. "Climate, Work and Labour: The International Context." In *Climate@Work*, ed. C. Lipsig-Mummé, 21–40. Halifax: Fernwood Publishing.

MacGregor, S. 2006. *Beyond Mothering Earth: Ecological Citizenship and the Politics of Care*. Vancouver: University of British Columbia Press.

Mackenzie, H., H. Messinger, and R. Smith. 2008. *Size Matters: Canada's Ecological Footprint, by Income*. Toronto: Canadian Centre for Policy Alternatives. https://www.policyalternatives.ca/sites/default/files/uploads/publications/National_Office_Pubs/2008/Size_Matters_Canadas_Ecological_Footprint_By_Income.pdf.

McFarland, J. 2013. "The Gender Impact of Green Job Creation." Toronto: Work in a Warming World Research Project.

Metcalf, G.E. 2008. "Protecting the Poor with a Carbon Tax." Presentation to the Financing for Development Office, UN Department of Economic and Social Affairs, New York, 17 June.

Nelson, V., and T. Stathers. 2009. "Resilience, Power, Culture and Climate: A Case Study from Semi-Arid Tanzania, and New Research Directions." *Gender & Development* 17 (1):81–94.

OECD (Organisation for Co-operation and Development). 2008. *Household Behaviour and the Environment: Reviewing the Evidence*. Paris: OECD. http://www.oecd.org/greengrowth/consumption-innovation/42183878.pdf.

Posner, E.A., and C.R. Sunstein. 2008. "Global Warming and Social Justice." *Regulation* (Spring):14–20.

Rees, W.A. 2006. "Ecological Footprints and Bio-capacity: Essential Elements in Sustainability Assessment." In *Renewables-Based Technology: Sustainability Assessment*, ed. J. Dewulf and H. Van Langenhove, 143–58. Chichester, UK: John Wiley and Sons.

Riddell, J. 2011. "Montreal Meeting Shows Growing Support for Climate Justice." *Bullet*, Bulletin No. 493. 25 April.

Schultz, I., and I. Stiess. 2009. "Gender Aspects of Sustainable Consumption Strategies and Instruments." Frankfurt am Main: Institute for Social-Ecological Research.

Shipp, S. 1987. "Spending Patterns of Men and Women – Consumer Expenditure Survey Conference Paper Summaries." Proceedings of the American Statistical Association, Business and Economics Section. http://www.thefreelibrary.com/Spending+patterns+of+men+and+women.-a06712056.

Stahl, E., W. Rees, and M. Byers. 2011. *Achieving Climate Justice: A Framework for Action in a Climate of Chaos*. Vancouver: Canadian Centre for Policy Alternatives.

Statistics Canada. 2007. *Labour Force Survey*. Ottawa: Statistics Canada.

—. 2009. *Canadian Vehicle Survey: Annual*. Cat. No. 53-223-X. Ottawa: Statistics Canada.

—. 2010. "Travel Survey." *Labour Force Historical Review 2009*. Cat. No. 71F0004XVB. Ottawa: Statistics Canada.

—. 2012a. *Labour Force Survey*. Cat. No. 71-001-X. Ottawa: Statistics Canada.

—. 2012b. *North American Industry Classification System (NAICS) Canada*. Cat. no. 12-501-X. Ottawa: Statistics Canada. http://www.statcan.gc.ca/subjects-sujets/standard-norme/naics-scian/2012/index-indexe-eng.htm.

UNEP (United Nations Environment Programme). n.d. "Green Economy." http://www.unep.org/greeneconomy/.

US (United States). EPA (Environmental Protection Agency). 2014. "Environmental Justice." Last updated 23 September. http://www.epa.gov/environmentaljustice/.

Victor, D. 2011. *Global Warming Gridlock: Creating More Effective Strategies for Protecting the Planet*. Cambridge: Cambridge University Press.

Whiteside, H., and M.G. Cohen. 2012. "Gender and Climate Change – Bibliography (1990–2011)." http://warming.apps01.yorku.ca/wp-content/uploads/Gender-and-Climate-Change-Bibliography.pdf.

CHAPTER 4

CANADIAN LABOUR'S CLIMATE DILEMMA

GEOFFREY BICKERTON AND CARLA LIPSIG-MUMMÉ

> Creating a global economy that emits a half or a quarter of the greenhouse gases of today's economy will require wholesale changes in the way economic activities take place. (Fankhauser, Sehlleier, and Stern 2008, 427)

> Perhaps the most critical challenge to the trade union movement ... is to break out of its social confinement and isolation in order to intervene directly in the primary area of economic decision making. (Levinson 1972, 142)

INTRODUCTION

In the late spring of 2013, the planet crossed the dangerous threshold of 400 parts per million of carbon dioxide in the atmosphere (Kunzig 2013). Despite international efforts, the world is warming more rapidly than expected (Carey 2012). And as the complexity, destructiveness, and speed of changes to the climate increase, responding to climate change[1] has become an urgent social issue for all countries (Dupressoir et al. 2007). Canada is no exception. In the struggle to slow global warming, however, the world of work occupies a critical but neglected place as a major producer of greenhouse gases (GHGs). Climate change will be the prime shaper of how we work during the 21st century, and the jobs-versus-the-environment issue haunts the labour movement in every developed country.

In 2010, the Work in a Warming World (W3) research group began a five-year project funded by the Social Sciences and Humanities Research Council of Canada. Unions were partners and researchers, but the labour

Work in a Warming World, edited by Carla Lipsig-Mummé and Stephen McBride. Kingston: School of Policy Studies, Queen's University.

movement was not the focus of the research. In the course of W3's research, however, we came to understand how crucial labour was to crafting an evolving Canadian strategy to reduce the causes of climate change and the challenges that labour was facing in playing its role.

The first and second sections of the chapter frame the issues facing labour environmentalism in Canada's chilly climate. The third and fourth sections explore the evolution of labour thinking, analyzing more than 100 union documents and research projects. On research, we present the international survey carried out by the Canadian Union of Postal Workers (CUPW) on the range of actions taken by postal workers to reduce GHG emissions in their work practices, and we analyze it as a case study in the unexpected ways that research, carried out and shared through global union channels, can advance the greening of work. The conclusion asks, Would a movement-wide turn toward labour environmentalism, rethinking economic and political strategies, change the movement itself?

CONTEXT

In Canada, as elsewhere, neither the need for paid work nor rapid climate warming will disappear. Serious science has demonstrated that responding to climate change is necessary for the survival of the planet and society. Indeed, Ross Garnaut, Australia's leading climate economist, observes that global warming of more than 3 percent centigrade "would be associated to increased risks to the stability of state and society" (quoted in Christoff 2014, 142). Climate change is arguably the most important challenge to society, and to work and labour, that will be encountered in this century. But an effective response to global warming will entail a national policy for active labour market transitions as well as the redesign of work itself.

In the absence of policy to cushion the job loss caused by climate change and responses to climate change, the realistic fear of job loss will continue to slow or paralyze effective environmental action on the part of labour. While the decentralized structure of the Canadian industrial relations regime makes it difficult to negotiate at a national level for major investment in green technologies and a program of active labour market transition for workers affected by decarbonization, the labour movement already has the capacity to adapt work and workplaces in order to significantly reduce the carbon emissions produced by work. Developing strategies to slow global warming and green work are intimately linked.

The direct and flow-on impacts of climate change affect the operation of unions as well as the provision of jobs and the organization of the labour process. Climate's warming also affects workers' lives and futures more broadly. In both prosperous and poor countries, climate-induced exile, climate migration, coastal-community destruction, climate-induced precarity, food and water insecurity, and the growth of climate-created health problems are threat multipliers, widening the gap of vulnerability between classes, creating new tensions. At the same time, the continuing deindustrialization of the first industrialized countries and the industrialization of developing countries are triggering small and large green industrial revolutions.

Almost alone among industrialized countries, Canada has no national strategy to respond to climate change and is unlikely to meet its Kyoto target (Sustainable Prosperity 2012). To say that the Conservative government has been reluctant to acknowledge the reality of global warming is a generous understatement (Billett and Bowerman 2009). Since 2011, the government has withdrawn from the Kyoto Protocol as well as the United Nations desertification program. In a recent survey of climate and energy legislation in 66 countries, Canada ranked among the worst four (GLOBE International 2014).[2]

For this Conservative government, environmental science itself has come to be seen as an enemy. But the government's efforts to block public knowledge of the results of environmental research have muzzled climate scientists even more comprehensively than the United States experienced during the Bush era (*Economist* 2013; NUPGE 2013). The government's success in blocking research that threatens the expansion of the fossil fuel industry is part and parcel of the larger agenda to return to a resource-driven economy. The collateral damage of this return to a resource-based economy for the public services and employment in manufacturing is stark (Garmulewicz 2012). Since 2000, the Canadian Auto Workers (CAW; now UNIFOR)[3] has named the loss of one-third of the country's manufacturing jobs "resource-driven deindustrialization" (Stanford 2013).

Political life in Canada has also been compromised. In a recent radio interview with the Australian Broadcasting Corporation, Kevin Taft, former leader of the Liberal Party in Alberta, argued that democratic government in Canada had been captured by the carbon industry (Taft 2014). In a recent *New York Times* op-ed piece, Jacques Leslie echoed that concern (2014). The government has demoted the environment portfolio in Cabinet and reduced the number of governmental agencies carrying out environmental audits from 40 to 3 (Ross 2013). It has cut back, closed, or de-funded climate, oceans, and fisheries research centres as well as environmental research programs across the country (Davidson 2012). "Vanishing Science," a summary of two surveys by Environics Research Group commissioned by the Professional Institute of the Public Service of Canada directed at Canadians and federal scientists, found that "nearly three-quarters of Canadians (73%) believe the top priority for government scientific activity should be the protection of public health, safety and the environment" (PIPSC 2014, 7). Yet

> over three-quarters of federal scientists (78%) report cuts to capacity in their own workplace.... Nearly 7 out of 10 scientists (69%) at Environment

> Canada believe Canada is doing a worse job of environmental protection and sustainable resource management than five years ago. Over 8 out of 10 (83%) in the Department of Fisheries and Oceans believe the same. (ibid., 5)

The federal government has further weakened environmental oversight and national capacity to respond to climate change by gutting Environment Canada. "Although almost all [Environment Canada] programs are targeted for reduction, the 69 percent cut to climate programs between now and 2016–17 speaks volumes" (W3 2014a). Between now and 2016–17, the Conservative government will have fired more than 1,000 environmental scientists (Boutilier 2014).

Environmental literacy, skills training, and labour market research have also come under the gun. In 2011, the federal government announced that, in 2013, every national industry sector council in the country, key providers of occupational green training, would be de-funded (Buzzetti 2011). In a more obvious move, in 2013 the government closed the National Round Table on the Environment and the Economy (NRTEE), a highly respected advisory body established by a Conservative government in 1987. And other environmental think-tanks are undergoing tax audits to decide whether they will lose their charitable status. Decidedly, the federal government is willing to go an exceptional distance to muzzle inconvenient science.

Framing the Problem

In the Canadian struggle to slow global warming, the world of work is underdeveloped terrain. But work matters to climate change, and climate change matters seriously to the future of work. On one hand, climate change is becoming the most important force in determining how and where we work. On the other, the recent study by the International Labour Organization (ILO) on the European Union (EU) found that work creates 80 percent of the GHG emissions produced by human activity: work is at once a major producer of GHGs and a potentially crucial site for reducing them (ILO 2011).

In the EU and the Scandinavian countries, global warming is taken as a whole-of-society threat. In these countries, policy and action to reduce the GHG emissions produced by work are led by governments, with unions, scientific organizations, environmental and labour market think-tanks, business, and charitable foundations collaborating in the research, policy development, industrial, and educational spheres. Not so in Canada, where multi-level, multi-actor, coordinated strategies to reduce GHG emissions are not part of our national political vision.

But the unions have resources for entering the climate struggle that draw on a uniquely Canadian range of strengths. In terms of action, unions have reach, discipline, and legal rights to introduce climate issues that affect health and safety. In terms of voice, the union movement is Canada's largest member organization in civil society, and it maintains collective bargaining relationships with thousands of employers. In terms of presence, union density remains relatively stable, around 31 percent, so unions maintain a role in societal decisions. In terms of legal protection, long-established federal and provincial laws maintain workplace health and safety laws, as well as collective bargaining rights, that allow climate bargaining – collective negotiation for environmentally responsible labour processes. Union action to win and protect these laws has established a culture of continuing worker education that is now coming to include environmental literacy (Corbyn and Mann 2008). And unions, particularly in the public sector, can and do climate bargain (cf. CUPE 2006, 2008). Finally, Canadian workplaces, unionized or not, operate in a highly regulated environment.

Notwithstanding the federal government's climate silence, Canadians are concerned about global warming. In a survey carried out by Canada 2020 on American and Canadian priorities, for example, 77 percent of Canadians were very concerned or somewhat concerned about climate change, while 57 percent of Americans were very or somewhat concerned (W3 2014c). In ranking the importance of national policy priorities, Canadians ranked protecting the environment number one.

The combination of public concern about global warming and government silence poses both threat and opportunity to the labour movement. The opportunity is to fill the existing national void in climate leadership. The threat looks like this: the federal government's vision of returning Canada to a fossil fuel–based resource economy is widening social inequality and speeding the degradation of the environment. At the federal level, labour's voice has been difficult to hear. The high-dollar effect is creating resource-based deindustrialization, which condemns manufacturing to a continued decline in jobs, the private sector to continuing growth in precarious employment, and the resource industries to exponential profits, with few ongoing jobs. The decline of manufacturing jobs and the precarization of jobs threaten all unions.

There is also an industrial relations component. At least since 2011, the federal government has been explicit about its goal to reduce wages and job protection and weaken union rights in collective bargaining and workplace representation in the public sector. This campaign, referencing the new right-to-work legislation in some American states, is now being taken up by provincial Conservatives and expanded to include parts of the private sector as well as pension and wage regimes. This threat, at the top of the list for Canada's unions, crowds out other urgencies.

Canada's political decentralization and the absence of a Canadian national climate plan that integrates economy and environment means that workers and their unions cannot assume that any future policy leading to a low-carbon transition will include the provision of or funding for adaptive green-skills training, financial support for workers who lose their jobs, regional economic planning for industry-shift, government support for green start-ups, or active labour market transitions. In this climate, the old-jobs-versus-environment argument re-emerges with intensity.

> The measures to enable the European Union to reduce its CO_2 emissions by some 40% by 2030 do not globally destroy jobs, but they do substantially change the supply and demand of jobs and qualifications within and between sectors.... Job movements are likely to take place in all sectors. (Dupressoir et al. 2007, 8)

Cushioning job churning and job loss has been critical for obtaining union buy-in to significant carbon reduction in the European work world (Dupressoir et al. 2007). It requires policy co-operation among all levels of government, environmental policy integrated with social and economic policy, active buy-in from business and labour, multi-level economic planning, and sectoral and regional support for active labour market transitions. In one of its last reports, the NRTEE asserted that national government leadership and coordination of the production of GHGs are essential (NRTEE 2012). With Canadian public policy based as it is on "the market knows best," and without the cushion that exists in Europe, it is not unreasonable for Canadian unions and their members to fear that any action to slow global warming without having programs in place for worker adaptation will cost jobs in brown industries without providing compensatory protection or transition to new work.

However, unexpected voices are now speaking out for environmental responsibility. Listen to Jeffrey Seabright, former vice-president for environment and water resources at the Coca-Cola Company, speaking at the World Economic Forum at Davos in early 2014. "Increased droughts, more unpredictable variability, 100-year floods every two years.... When we look at our most essential ingredients, we see those events as threats" (quoted in Davenport 2014). And according to Nike, drought is worsening in countries that produce the cotton it uses. Hannah Jones, Nike's vice-president of sustainable business and innovation, says, "That puts less cotton on the market, the price goes up, and you have market volatility" (quoted in Davenport 2014).

In addition, some of the largest American multinational corporations that also operate in Canada, and American manufacturing and construction-business associations, have begun to draw attention to the chaotic impact of climate change and wild weather on their supply chains. They are lobbying governments to pay attention to the impacts of global warming (Davenport 2014).

The Complex Map of Canadian Union Representation

In 2013, Canadian union density was 31.5 percent (Statistics Canada 2013). The movement's internal composition has been undergoing important changes in demographic composition, occupation, and job security for a generation, but Canada has not experienced the radical decline in union membership seen in other countries in the Organisation for Economic Co-operation and Development. Notwithstanding relatively robust union density, the Canadian manufacturing sector has lost almost a third of its labour force in the past decade, with vehicle production losing 30 percent of jobs in 2009 alone (Stanford 2010, quoted in Holmes and Hracs 2013). Forestry products, a leading Canadian industry, lost 40.6 percent of its jobs between 2004 and 2011 (Holmes 2013). As labour force participation by men declined after the mid-1970s, so too did male union membership as a proportion of total union membership. From the late 1970s, women's labour force participation grew steadily, both in absolute numbers and in comparison to male union membership (Lipsig-Mummé and Laxer 1999). Avoiding radical decline in union density in Canada, then, seems to be partly the product of the growth of women's employment and unionization.

Canada has not suffered the hollowing out of unions' legal rights that some American states are experiencing, although the same shift may be starting to happen. For example, the Canadian government has signalled that high on its list of priorities is cutting union "privileges." Some provinces are borrowing tactics from the newest American assault on unions, and New Brunswick's program of replacing secure public sector pensions with shared risk has been taken up in the United States and is under consideration by several provinces and the federal government for its public sector workers. The decentralization of the Canadian industrial relations regime and labour's fading voice in national policy now make it very difficult for labour to influence federal industrial-relations or training policy and active labour market transitions at the national level. For workers whose brown industries will scale down or retool eventually, the weakness of the labour voice at the national level is a pressing anxiety. Ironically, however, as industrial relations decentralization contributes to labour weakness in setting national policy, it strengthens the unions' local voice and power, creating unexpected opportunities for labour environmental action at the level of the workplace, the industry, the province, and the municipality.

The fact that union density hovers around 30 percent should be a source of unusual strength in this time of generally shrinking union membership in the Global North. Sadly, it is not. As well as being affected by the decline of manufacturing, Canadian labour is also weakened by fragmentation and an exceptional degree of overlap and duplication among unions.

The growth of general unions by consolidating or absorbing sectoral and occupational unions that straddle the public and private sectors also makes it harder for union representatives to grasp the labour processes of their members' diverse work. A general union's decision to bring climate issues into the workplace presents its staffers and officers with the challenge of crafting a climate plan for a very wide range of occupations and work organization. Not surprisingly, many general unions have produced compelling statements of principle concerning the environment but have difficulty operationalizing a plan of climate action for their varied membership.

Labour's Dilemma and Union Response

The preceding sections have outlined some of the obstacles confronting the Canadian world of work and labour as it grapples with global warming. In this section, we draw on more than 100 union documents[4] to explore the development of the union movement's response to the climate crisis. Produced between 1993 and the present, these documents are varied – collaborative how-to projects linking labour, management, nongovernmental organizations (NGOs), union policy statements, statements of principle, green training documents, educational materials, guides for climate bargaining, and research. They have been produced by national and local unions, provincial federations of labour, labour councils, and the Canadian Labour Congress (CLC), la Confédération des syndicats nationaux, la Fédération des travailleurs et travailleuses du Québec, and la Centrale des syndicats du Québec. Some of the documents have been authored collaboratively by multiple unions, by unions and community organizations, by unions and academics, and, more rarely, by collaboration between a labour body and quasi-governmental organizations. The documents discussed in this section are drawn from these 100-plus documents and were chosen for analysis because of their significance in the development of labour's environmental vision.

The development and fragmentation of labour's response to global warming is of particular interest, as is the evolution of the movement's definition of the limits and possibilities of its own role. So are the broken links: between statements of principle and action and results, between national policy and local action, between seeking solutions in the political realm and seeking them in the economic realm.

From the early 1990s, the Canadian union movement at the national level combined a willingness to collaborate with groups and governments of all political hues with fully developed ideas about the need for an autonomous role for unions in creating a sustainable society that was both low carbon and socially equitable. In 1993, for example, the CLC and NRTEE jointly produced *Sustainable Development: Getting There from Here – A Handbook for Union Environment Committees and Joint Labour-Management*

Environment Committees. In the preface, Bob White, then president of the CLC, and Dr. George Campbell, then chair of NRTEE, described the purpose of the handbook.

> In many ways, workers were the first environmentalists in the age-old struggle against workplace pollution and the pollution of communities adjacent to the workplace.... The aim of sustainable development has to be two-fold: to reconcile environmental protection with the well being and prosperity of society, and to enable workers to be full participants with other stakeholders in shaping the required societal change....
>
> This manual will help to ensure that workers will have a prominent role in advancement of sustainable development as far as possible through cooperation with environmentalists, governments, and employers.
>
> The collaboration of the Canadian Labour Congress and the National Round Table in this important publication should convey a strong message that the two organizations share a commitment to sustainability, and will pursue that commitment through cooperation wherever possible. (Schrecker et al. 1993, 5)

In 1993, this collaboration between the Canadian Labour Congress and an arm's-length advisory body created by a Conservative government was unremarkable and considered valuable.

In 1999, "Just Transition for Workers during Environmental Change" was adopted by the CLC (2000). It was a remarkable, trail-blazing, social climate-change agenda, placing workers and their unions in a pivotal role among the groups in society that should share the burden for greening the economy. (At that time in labour environmental activism, the terms *environment* and *climate change* were used interchangeably.) "Just Transition" saw labour collaborating with other groups to work for a whole-of-society solution, but it stressed that labour needed to articulate its independent vision of a sustainable society and develop it collectively within the labour movement. Central to the position paper was the belief that jobs and the greening of the economy should grow together, but that a green turn could not take place without an aggressive green development strategy, which the unions would play a critical role in developing. Greening the Canadian economy, then, was a process as much as a goal. Environmental adaptation of the economy was essential, but workers were already paying for environmental adaptation and would therefore be unwilling to shoulder costs that should be shared by many social groups.

Perhaps most striking about the document was the argument that the labour movement would not allow the idea of sustainable transition to be hijacked by capitalists, the state, or the environmental movement, although it would work with all. Fast-forward to the present, and so much has changed that a negotiator for one of the large industrial unions (who prefers to remain anonymous) has said, "The most we get [in terms of joint action to reduce GHG emissions] is the employer telling us what new equipment they're bringing in, and offering us some training" (Pers. comm. 2012).

Today, the "Just Transition" document of 2000 reads as a creative and organizationally self-assured project of society. It is a framework for national action that is fundamentally political, starting from the assertion that labour will play a major role in major social decisions. While "Just Transition" has since become an international byword for Canadian pioneering environmental leadership, at home very little of its ambitious vision has been put into practice. Over time, it has withered. By 2008, the updated version had narrowed its focus to obtaining adequate financial compensation for workers who lose their jobs because of climate change or the policies that respond to climate change (CLC 2008).

In the wake of "Just Transition," many Canadian unions produced labour-environmental policy statements for their websites. For some, the commitment to Planet Earth is at the heart of the matter; for others, it remains a linguistic bow. For still others, greening the economy is a dynamic social project that requires an industry strategy: environmental and labour market policy have to work in tandem. But in sectors where jobs are menaced by the lack of national policy to transition brown jobs to green work, the jobs-versus-the-environment stalemate remains.

The Alberta Federation of Labour (AFL) was typical of other labour federations when, in its 2007 "Climate Change Policy Paper," it called for a shift in focus from the environment to the more serious threat of climate change (AFL 2007). The shift in language signalled a sharpened focus. The AFL placed itself in the role of steward, charged with building climate awareness and activism in its province. Among the actions the AFL pledged to take were educating union members about the seriousness of climate change, producing Alberta-specific educational materials, launching a public education campaign to put labour's positions on climate change before the largest number of people, and including environmental education in its events. But the position paper did not discuss setting specific targets or developing reporting mechanisms to reduce GHG emissions in unionized workplaces.

During the global financial crisis and its prolonged restructuring, however, the labour movement was forced to focus on its struggle against the impact of austerity programs. Labour shifted its climate focus to green job creation, which became, in most provinces, its favoured defensive strategy in bad times (Good Jobs for All Coalition 2009). Despite the lack of a clear definition of a *green job* (Lipsig-Mummé 2011), the focus on green job creation served three purposes for the movement. Arguing that a green turn in the economy would create jobs, green job creation was meant to neutralize workers' fear of losing their jobs as a result of government regulation to reduce the causes of climate change (Cartwright 2011; Good Jobs for All Coalition 2009). The call for good green jobs was also meant to pressure governments to invest in renewable energy, infrastructure renewal, adaptive training, and other job creators (Green Jobs BC 2013). Finally, wide-ranging policies such as Ontario's *Green Energy and Green Economy Act* would lead to the revitalization of communities by growing clean-energy jobs at home, and it would rejuvenate domestic manufacturing by attracting investment in next-generation as well as traditional manufacturing.

Public policy, however, is vulnerable to politics – governments change, and parties in office shift their priorities. This has certainly been the case for Ontario, British Columbia, and New Brunswick where climate and environmental policy are concerned. In the years following 2011 and the formation of a majority federal Conservative government, the national union voice on climate change was less often heard, and a spectrum of union climate positions crystallized at all levels of the movement.

At one pole on the spectrum is the position that responding to climate change is the government's job: government must lead and fund the struggle to slow global warming, while the union's job is to push the government. Thus, in "Confronting Climate Change: A Just Transition to Green Jobs," a short document put out by the National Union of Public and General Employees (NUPGE) national office, the full list of what-you-can-do strategies consists of calling your municipal councillor, writing the premier, contacting your member of Parliament, and sharing this leaflet (NUPGE n.d.). Another short NUPGE document, "Get Our Governments to Go Green NOW!," lists the five top actions our governments must take: create national energy efficiency standards, build a green power grid, slow oil sands development, assure a just transition for displaced workers to new green jobs, and set true hard caps on GHG emissions (NUPGE 2013). From this perspective, the union is to act as government goad, organizer of civil society pressure on the government to do the right thing for global warming, and idea-generator for the government's priorities.

At the opposite pole of the spectrum is the position that workers and their unions can begin by greening their environment at work and in the wider society. To that end, the Canadian Union of Public Employees (CUPE) adopted a 2013 national environmental policy entitled "Working Harmoniously on the Earth" (CUPE 2013). CUPE takes the position that our workplace is our environment. While government policy is important, CUPE focuses on what workers themselves can do, from carrying out research to climate bargaining to forming broader alliances. Linking policy statement to education and action, the national union widely distributes a series of action manuals for forming a workplace environment committee: "You Can Help Your Workplace Go Green" (CUPE 2012b),

"Green Bargaining for CUPE Locals" (CUPE 2008), "A Worker's Action Guide to a Greener Workplace" (CUPE 2012a), "How to Elect a Green Delegate" (CUPE 2006), and "Eco-audit Your Workplace" (CUPE n.d.).

Across the country, environment committees are increasingly common in unionized workplaces and at the national level in unions. They may be informal groups. Or they form within occupational health and safety committees, and remain there, or become an independent committee. In all cases, the issues that environment committees take up are constantly expanding to include health, environmental literacy, green retraining, collective bargaining, and the relations between plant and community in terms of environmental health. Some examples: UNIFOR has local and national environment committees that link with youth outreach groups (CAW 2013). The Confédération des syndicats nationaux produces a handbook for their youth conferences and a union guide on the environment (CSN 2011). CUPE has more than 100 functioning workplace environment committees.

Unions have also begun to use collective bargaining to link greening the workplace to greening the product or the labour process. For example, CUPW has negotiated with its principal employer, Canada Post, on greening the vehicle fleet, the largest in Canada (CUPW 2010). The CAW has carried out influential research on green vehicles, extended producer responsibility, and fuel efficiency standards (Hargrove 2004). Unions have set up training courses and produced training material for hotel cleaners in Toronto – not so much greening their product as greening the toxic tools of their trade (Labour Environmental Alliance Society 2008).

The examples above indicate that responding to environmental upheavals as well as greening the process of production will require new or adaptive-skills training. In a widening range of outdoor and indoor jobs, wild weather and global warming's rising temperatures now require not only new skills training but also protective clothing and disaster training. These climate issues are starting to appear on the collective bargaining radar for Canadian unions.

However, unions in heavy industry, facing major job closures or job churning if their brown industries go green without national, active job-transition strategies in place – or those that have already faced this upheaval – must walk a tightrope where environmental responsibility is concerned. They remain cautious about the threats to jobs from greening work or adaptive retraining. They may support government plans that boost employment in environmentally sensitive industries and communities, even when this puts them on a collision course with environmental groups and other unions.

The forestry industry is a case in point. Faced with a major global downturn in demand, coupled with a pine beetle devastation caused by warming temperatures, between 2004 and 2009 the Canadian forestry industry lost 40.6 percent of its workforce (Holmes 2013). In 2014, the

New Brunswick Progressive Conservative government unveiled its first forestry plan in 25 years. The plan outlined the government's decision to allocate substantial additional hectares of Crown land to industry, and it has attracted fierce opposition from scientists and environmentalists. However, it is supported by UNIFOR, the largest union representing forestry workers in the province. Jerry Dias, its president, had this to say: "We are very happy to see this new plan. This is a perfect example of government, companies and organized labour working together for the betterment of our economy" (UNIFOR 2013).

Overall, then, what do we know now about the effectiveness of the active measures proposed by labour to reduce GHG emissions and conserve jobs? We do not have concrete answers, but "Green Bargaining for CUPE Locals" (CUPE 2008) sets out some examples of success. Among them: negotiating a workplace environmental policy between a Saskatchewan local and an NGO that includes "an extensive action plan for sustainable purchasing, energy conservation, waste reduction, green meetings, and green transportation options" (3). In Mount Pearl, Newfoundland, the city and a CUPE local negotiated "that the parties affirm ... their joint objective to co-operate and promote jointly the objective of a pollution free environment at work and in the community" (3). Sometimes a union committee can operate informally and get results. In Niagara, Ontario, CUPE members working for the district school board set up an informal energy conservation committee, which "helped to save more than five million kilowatt hours of power" (7).

The New Brunswick Union of Public and Private Employees (NBU), a component of NUPGE, has perhaps gone further than other Canadian unions in developing a strategy to reduce GHG emissions and protect jobs. NBU has developed bargaining language that calls for "both parties to reduce the carbon footprint of the workplace by 3% per year over the duration of this agreement.... The savings shall be distributed on a 50%-50% basis between the employer and the Union for distribution on further greening processes" (quoted in CUPE 2008, 6). NBU has also bargained for the establishment of an energy conservation committee to reduce GHG emissions as well as model language to negotiate a Just Transition agreement by enterprise or workplace (9). Finally, the union has published *Cool Comforts: Bargaining for Our Survival – A Union Activist's Handbook on Global Warming* (Corbyn and Mann 2008).

It must be said that the examples discussed above are promising steps forward, but, except for a few cases, they do not demonstrate a reduction of GHG emissions or labour's climate leadership. For Canadian unions, the challenge is to link effective modes of action at the local, municipal, and provincial levels with action and influence at the national level. At present, these are two solitudes. There is little strategic cross-pollination between national-level labour and provincial- and local-level labour-climate work. Labour's lack of influence in the national policy arena has made it difficult to push for a national climate policy, and it has made it impossible to push for the creation of the active labour market transitions that are essential if workers in threatened brown industries are to buy in to the necessity to green work.

Case Study: CUPW's International Research

Research, however, is now being used to broaden the capacity of some Canadian unions to engage their members on climate issues. Unlike European unions, which have established think-tanks to inform their policy and action, Canadian unions draw on the work of academics on an ad hoc basis. As an example, the enormous disruptions that climate change is creating and will create for society and work have led CUPW to use research more ambitiously. Its long-term involvement with a multi-partner research project on climate change and the future of work, and the refractory nature of its relationship with its major employer, has led it to engage its members in a recent international survey on the state of environmental knowledge and action in postal unions around the globe.

CUPW represents approximately 55,000 people, with urban operations accounting for 48,000 employees and rural/suburban operations constituting 6,600 employees. CUPW includes all of the operational workers at Canada Post as well as couriers and sorters at approximately 15 private-sector courier companies. Historically, the union has viewed itself as the guardian of the public postal service in Canada, with a mandate to ensure the delivery of high-quality postal services geared to meet the needs of working people. To this end, the union has worked with coalition allies to rigorously resist any attempts to deregulate or privatize postal services. The union has also fought against free trade agreements designed to restrict public services and enhance the role of the private sector in the postal and courier sectors.

More recently, the union acknowledged that it has a responsibility to ensure that the postal and courier services provided by its members are performed in an environmentally responsible manner. CUPW recognized that with its primary employer having the largest vehicle fleet and the largest number of retail facilities of any employer in the nation, it had a special responsibility to address environmental issues in a labour relations context. In 2008, CUPW joined several other major unions, academics, and environmental activists in W3. One of the union's first projects was to conduct an analysis of the GHG footprint of the postal and courier sector.

In its 2008 submission to the strategic review of Canada Post Corporation (CUPW 2008), CUPW made several recommendations concerning the need to reduce GHG emissions in the sector. It recommended that the federal government should sponsor a thorough examination of the overall environmental impact of all postal and courier services, including an environmental assessment of the different delivery modes, such as door-to-door delivery and community mailboxes. The union also proposed that such a review should examine how the industry could be reorganized to operate in a more environmentally friendly manner. In addition, CUPW called on Canada Post to conduct an environmental audit to identify measures that could be taken to reduce its carbon footprint.

In its submission, the union argued that greater competition in letter delivery, as advocated by various right-wing think-tanks, would create more environmental problems: there is a direct and inverse relationship between increased delivery density and environmental impact as the decreased delivery density created by competition would lead to an increased use of fossil fuels, pollution, and traffic. According to the union, from an environmental perspective, it makes sense not only to maintain the letter monopoly but also to extend it to parcel delivery. Moreover, the union asserted that the postal service can and should be used to develop and test environmental practices that could be extended to other industries.

In 2010, partially in response to the positive reception of the results of the W3 research, CUPW developed and conducted three-day educational programs on postal services and the environment in several regions. At its 2011 national convention, the union adopted a comprehensive policy on the environment, calling on its members to work with environmental activists and to address the GHG footprint in the postal sector. In 2013, the union decided to establish a national network of video conferencing, which has significantly reduced the amount of airline travel taken by union representatives. Also in 2011, for the first time, CUPW entered into its national negotiations with Canada Post with a proposal to include a new article in the national collective agreement that would require the parties to work together to reduce the corporation's environmental footprint.

Throughout the negotiations, management stiffly opposed all of these proposals. In the end, the negotiations were not successful, and the union commenced rotating strikes in June 2011. Almost two weeks later, management responded with a national lockout, and two weeks after that, Parliament passed special back-to-work legislation, imposing a process of mandatory arbitration. The final settlement of the arbitration, reached in December 2012, includes only a letter of agreement entitled "Environmental Initiatives," which is limited to testing the feasibility of a recycling service and reviewing how both parties can reduce their use of paper.

Following the negotiations, the union decided to proceed unilaterally to investigate the best practices of other postal unions concerning the environment. Together with W3 and UNI (Union Network International)

Global Union, the international federation with which CUPW is affiliated, it developed and distributed a survey to all postal unions around the globe.

The survey, originally produced in English, Spanish, and French and later translated into Japanese, asked the unions what they were doing to reduce their own GHG emissions as well as those of their employers. It asked them to provide any policies or educational materials they had produced. It also asked about any modifications in work processes, equipment, or uniforms that had been introduced and whether the unions had been involved in the decision-making that had led to the changes. Finally, in the context of worker rights, the survey asked whether the unions had attempted to negotiate collective agreement clauses that addressed the impact of climate change.

The impact of the survey on the unions involved around the globe continues to spread. After participating in the international W3 conference in late 2013, UNI Global Union decided to make climate change a priority for its 2014 World Congress.

CONCLUSION

At the beginning of this chapter, we referred to the dilemma that the climate crisis poses to the Canadian labour movement. We might also argue that the void created by a lack of climate policy in Canada poses a threat, an opportunity, and a critical space to be filled.

Can labour occupy it? The labour movement does not contest the urgency of reducing GHG emissions. But neither has it taken in, or taken up, the real magnitude of the crisis. In other words, environmental action is not yet a wave in the Canadian labour movement. Labour environmental action is not yet on the same level as the big issues that lead unions to transform into social movements, drawing in increasing numbers of ordinary members of every age, who bring their creative ideas; generating strategy at each level of union life; sharing ideas among unions, provinces, and age groups who usually do not meet; brainstorming with community organizations, environmental groups, and the global union federations; developing new tactics and strategies; and using them.

Making climate change a real priority requires deep changes in today's fragmented Canadian labour movement. First off, labour's traditional economic focus and traditional political tactics will need to change. As early as 1972, Charles Levinson, the visionary thinker of the international labour movement, called on labour "to break out of its social confinement and isolation in order to intervene directly in the primary area of economic decision making" (142). The need for labour to intervene economically in a globalizing world led Levinson to develop a cross-cutting strategy: First, coordinate collective bargaining by all unions across the globe representing workers in each of the large multinational corporations. Second, coordinate all unions in each of the major industrial sectors into sectoral world company councils. Coordinated collective bargaining would become transnational framework agreements. Today's global union federations play some of the roles envisaged for the world company councils.

If responding to climate change becomes a serious priority for Canadian labour, creatively adapting Levinson's vision and strategy to the climate crisis of the 21st century might open paths that are currently blocked. Here are three examples. Focusing nationally rather than internationally, all unions in each of the major economic sectors – the food industry, health industry, auto industry, etc. – could collectively negotiate Canadian-based sectoral agreements on environmental responsibility. These agreements would make provision for locally negotiated climate and environmental clauses. Second, unions could negotiate global agreements on environmental responsibility with multinational companies headquartered in Canada, with the potential to broaden these agreements to take in the companies' international operations and collaborate with unions in those countries. Third, focusing nationally or, if need be, provincially, unions could propose legislation that imposes environmental responsibility on all Canadian-based companies and their overseas dependents.

In a globalizing world, rethinking the way unions can influence the primary areas of economic decision-making, and the legal environmental responsibility of employers, would be an important step forward.

NOTES

1. *Climate change* is "a change of climate which is attributed directly or indirectly to human activity that alters the composition of the global atmosphere and which is in addition to natural climate variability observed over comparable time periods" (UNFCCC 2013, para. 2). "It is *extremely likely* that human influence has been the dominant cause of observed warming since the mid-20th century" (IPCC 2013, 17).
2. The survey is also the subject of an article in the *Work and Climate Change Report*, a monthly publication of W3.
3. UNIFOR was formed on 31 August 2013 as a merger of the Canadian Auto Workers and the Communications, Energy and Paperworkers Union of Canada (CEP). It is Canada's largest private-sector union, with over 305,000 members.
4. The documents consulted are "grey literature," documents produced outside scholarly journals and mainstream commercial publishing (Perry 2013); it is notoriously evanescent (ibid.). But union grey literature – distributed at meetings; sent out by a union's national office to its locals; reports on union research, prepared for training sessions or collective bargaining, posted on a website for a period of time, etc. – also offers unique insights into the interplay between national and local, formal and informal, in the growth of labour's climate activism.

REFERENCES

AFL (Alberta Federation of Labour). 2007. "Climate Change Policy Paper." 45th Constitutional Convention, May.

Billett, S., and N. Bowerman, eds. 2009. *Assessing National Climate Policy: November 2008–February 2009*. Climatico. http://www.climaticoanalysis.org/press-releases/Climatico_National_Assessment_Report_March2009.pdf.

Boutilier, A. 2014. "Environment Canada Braces for Cuts to Climate Programs." *Thestar.com*, 12 March. http://www.thestar.com/news/canada/2014/03/12/environment_canada_braces_for_belttightening.html.

Buzzetti, H. 2011. "Ressources humaines – Ottawa coupe en catimini." *Le Devoir*, 19 July. http://media1.ledevoir.com/politique/canada/327711/ressources-humaines-ottawa-coupe-en-catimini.

Carey, J. 2012. "Is Global Warming Happening Faster Than Expected?" *Scientific American*, November:51–55.

Cartwright, J. 2011. "Green Jobs Are the Future." Presentation at the "Green/ing Jobs: Definitions, Dilemmas, Strategies" panel event, Toronto, 20 January. http://warming.apps01.yorku.ca/wp-content/uploads/2011/08/Green-Jobs-final.pdf.

CAW (Canadian Auto Workers). 2013. "CAW Health, Safety and Environment Newsletter."

Christoff, P., ed. 2014. *Four Degrees of Global Warming: Australia in a Hot World*. Abingdon, UK: Earthscan / Routledge.

CLC (Canadian Labour Congress). 2000. *Just Transition for Workers during Environmental Change*. Ottawa: CLC.

—. 2008. "Climate Change and Green Jobs: Labour's Challenges and Opportunities." Document 9 (as Amended) of the Proceedings of the 25th CLC Constitutional Convention, Toronto, 26–30 May.

Corbyn, P., and T. Mann. 2008. *Cool Comforts: Bargaining for Our Survival – A Union Activist's Handbook on Global Warming*. Ottawa: National Union of Public and General Employees.

CSN (Confédération des syndicats nationaux). 2011. *Des gestes pour l'avenir: Guide syndical en environnement*. Montreal: CSN.

CUPE (Canadian Union of Public Employees). 2006. "How to Elect a Green Delegate." Ottawa: CUPE.

—. 2008. "Green Bargaining for CUPE Locals." Ottawa: CUPE.

—. 2012a. "A Worker's Action Guide to a Greener Workplace." Ottawa: CUPE.

—. 2012b. "You Can Help Your Workplace Go Green: How to Form an Environment Committee." Ottawa: CUPE.

—. 2013. "Working Harmoniously on the Earth: CUPE's National Environment Policy." Ottawa: CUPE. http://30.cupe.ca/files/2013/01/Working_harmoniously_on_the_Earth_FINAL.pdf.

—. n.d. "Eco-audit Your Workplace." Ottawa: CUPE.

CUPW (Canadian Union of Postal Workers). 2008. "Submission of the Canadian Union of Postal Workers to the Canada Post Corporation Strategic Review." Ottawa: CUPW.

—. 2010. "CUPW's Environment Vision Gives Results." Environmental bulletin, 2 December. Ottawa: CUPW.

Davenport, C. 2014. "Industry Awakens to Threat of Climate Change." *New York Times*, 23 January. http://www.nytimes.com/2014/01/24/science/earth/threat-to-bottom-line-spurs-action-on-climate.html?_r=0.

Davidson, A. 2012. "Ottawa to Slash Environment Review Role: Critics Accuse Tories of 'Abdicating' Government's Responsibility to Protect Environment." *CBC-news* online, 17 April. http://www.cbc.ca/news/politics/story/2012/04/17/environmental-reviews.html.

Dupressoir, S. et al. 2007. *Climate Change and Employment: Impact on Employment of Climate Change and CO_2 Emission Reduction Measures in the EU-25 to 2030 – Synthesis*. Brussels: European Trade Union Confederation / Social Development Agency.

Economist. 2013. "Scientific Freedom in Canada: Keep It to Yourselves." Americas View, 7 March. http://www.economist.com/blogs/americasview/2013/03/scientific-freedom-canada.

Fankhauser, S., F. Sehlleier, and N. Stern. 2008. "Climate Change, Innovation and Jobs." *Climate Policy* 8:421–29.

Garmulewicz, A. 2012. "Slashing of Agency Reveals Canadian Reliance on Outdated Economic Thinking." *Thestar.com*, 25 May. http://www.thestar.com/opinion/editorialopinion/2012/05/25/slashing_of_agency_reveals_canadian_reliance_on_outdated_economic_thinking.html.

GLOBE International. 2014. *The GLOBE Climate Legislation Study: A Review of Climate Change Legislation in 66 Countries*. 4th ed. London: GLOBE International / Grantham Research Institute. http://www.globeinternational.org/studies/legislation/climate.

Good Jobs for All Coalition. 2009. "Good Green Jobs for All: Framework for Action Developed from the November 2009 Conference." Toronto: Good Jobs For All Coalition.

Green Jobs BC. 2013. "Moving Towards a Bold Green Jobs Plan for BC." Vancouver: BCGEU.

Hargrove, B. 2004. "Re: Automotive Fuel Efficiency Standards." Letter to various federal cabinet ministers. 28 October.

Holmes, J. 2013. "The Forestry Industry." In *Climate@Work*, edited by C. Lipsig-Mummé, 124–40. Halifax, NS: Fernwood Publishing.

Holmes, J., and A. Hracs. 2013. "The Transportation Equipment Industry." In *Climate@Work*, edited by C. Lipsig-Mummé, 105–23. Halifax, NS: Fernwood Publishing.

ILO (International Labour Organization). 2011. *Towards a Greener Economy: The Social Dimensions*. Geneva: ILO / International Institute for Labour Studies.

IPCC (Intergovernmental Panel on Climate Change). 2013. "Summary for Policymakers." In *Climate Change 2013: The Physical Science Basis – Contribution of Working Group 1 to the Fifth Assessment Report of the Intergovernmental Panel on Climate Change*. Cambridge: Cambridge University Press. http://www.climatechange2013.org/images/report/WG1AR5_SPM_FINAL.pdf.

Kunzig, R. 2013. "Climate Milestone: Earth's CO_2 Level Passes 400 PPM." *National Geographic News*, 9 May. http://news.nationalgeographic.com/news/energy/2013/05/130510-earth-co2-milestone-400-ppm/.

Labour Environmental Alliance Society. 2008. "Training for a Non-toxic Workplace." Vancouver: Labour Environmental Alliance Society.

Leslie, J. 2014. "Is Canada Tarring Itself?" *New York Times*, 31 March. http://www.realclearpolitics.com/2014/03/31/is_canada_tarring_itself_328863.html.

Levinson, C. 1972. *International Trade Unionism*. London: Routledge.

Lipsig-Mummé, C. 2011."Greening Jobs and Work: Slowing Global Warming and the Canadian Economy." Toronto: Work in a Warming World.

Lipsig-Mummé, C., and K. Laxer. 1999. "Organising and Union Membership: A Canadian Profile in 1997." Ottawa: Canadian Labour Congress.

NRTEE (National Round Table on the Environment and the Economy). 2012. "Reality Check: The State of Climate Progress in Canada." Ottawa: NRTEE.

NUPGE (National Union of Public and General Employees). 2013. "Get Our Governments to Go Green NOW!" Flyer. 29 September.

—. n.d. "Confronting Climate Change: A Just Transition to Green Jobs." Ottawa: NUPGE. http://nupge.ca/sites/nupge.ca/files/publications/Environment/Confronting_Climate_Change.pdf.

Perry, E. 2013. "Changing Patterns in the Literature of Climate Change and Work." In *Climate@Work*, ed. C. Lipsig-Mummé, 11–19. Halifax, NS: Fernwood Publishing.

PIPSC (Professional Institute of the Public Service of Canada). 2014. "Vanishing Science: The Disappearance of Canadian Public Interest Science." Ottawa: PIPSC. http://www.pipsc.ca/portal/page/portal/website/issues/science/vanishingscience/pdfs/fullreport.en.pdf.

Ross, P. 2013. "Opinion: Canada's Mass Firing of Ocean Scientists Brings 'Silent Summer'." *Environmental Health News*. http://www.environmentalhealthnews.org/ehs/news/2012/opinion-mass-firing-of-canada2019s-ocean-scientists.

Schrecker, E., H. Mackenzie, J. O'Grady, and the CLC. 1993. *Sustainable Development: Getting There from Here – A Handbook for Union Environment Committees and Joint Labour-Management Environment Committees*. Ottawa: CLC / National Round Table on the Environment and the Economy.

Stanford, J. 2013. "Resource-Driven Deindustrialization." Submission to the House of Commons Standing Committee on Natural Resources on behalf of the CAW. 23 April.

Statistics Canada. 2013. "Labour Force Survey Estimates (LFS), Employees by Union Coverage, North American Industry Classification System (NAICS), Sex and Age Group, Annual." CANSIM Table 282-0078. Ottawa: Statistics Canada.

Sustainable Prosperity. 2012. "The United Kingdom (UK) Climate Policy: Lessons for Canada." Policy Brief. Ottawa: Sustainable Prosperity. http://www.sustainableprosperity.ca/dl821&display.

Taft, K. 2014. "Democracy 'Captured' by Carbon Industry in Canada: Taft." Interview by Ellen Fanning. *RN Breakfast*. Radio program, Australian Broadcasting Corporation, 21 March. http://www.abc.net.au/radionational/programs/breakfast/democracy-captured-by-carbon-industry-in-canada-taft/5335946.

UNFCCC (United Nations Framework Convention on Climate Change). 2013. "Full Text of the Convention: Article 1 – Definitions." http://unfccc.int/essential_background/convention/background/items/2536.php.

UNIFOR. 2013. "Atlantic's Largest Forestry Union Applauds Long-Term Plan for Sector." News release. 13 March. http://www.unifor.org/en/whats-new/press-room/atlantics-largest-forestry-union-applauds-long-term-plan-sector.

W3 (Work in a Warming World). 2014a. "Environment Canada's Cuts Diminish Our Capacity." *Work and Climate Change Report*. 26 March. http://www.workinawarmingworld.yorku.ca/projects/work-in-a-warming-world/.

—. 2014b. "Global Survey of National Climate and Energy Legislation Ranks Canada as a Laggard." *Work and Climate Change Report*. Posted by Ava Lightbody, 26 March. http://workandclimatechangereport.org/?s=Global+Survey+of+National+Climate+and+Energy+Legislation+&submit=Search.

—. 2014c. "Public Opinion on Climate Change Issues: North American and European." *Work and Climate Change Report*. Posted by Ava Lightbody, 26 March. http://workandclimatechangereport.org/2014/03/26/public-opinion-on-climate-change-issues-north-american-and-european/.

CHAPTER 5

RENEWABLE ENERGY DEVELOPMENT AS INDUSTRIAL STRATEGY: THE CASE OF ONTARIO'S *GREEN ENERGY AND GREEN ECONOMY ACT*

MARK WINFIELD

INTRODUCTION

The potential contributions of low-impact, renewable energy sources to climate change mitigation (IPCC 2011) and advancing energy sustainability (Winfield et al. 2010) have prompted many jurisdictions in the European Union (EU) and the United States to undertake initiatives to promote the development of renewable energy sources. These renewable energy initiatives have typically employed feed-in-tariff (FIT) programs, renewables obligations, or renewable portfolio standards. FIT programs provide renewable energy developers with a fixed, long-term price for the energy they produce and typically guarantee grid access as well. They have been adopted by many European jurisdictions, including Germany, Denmark, and Spain (Jacob 2012), as well as by Ontario. Renewables obligations or renewable portfolio standards require electricity suppliers to provide a set portion of their output from renewable sources. They have been employed in the United Kingdom (Edge 2006) and most US states (US Energy Information Administration 2012).

The goals of these renewable energy initiatives have generally gone beyond providing electricity supplies with lower environmental impacts and energy security risks than conventional, non-renewable alternatives.

Work in a Warming World, edited by Carla Lipsig-Mummé and Stephen McBride. Kingston: School of Policy Studies, Queen's University.

Rather, reflecting an evolution in the role of environmental technologies, renewable energy strategies are also conceived of as industrial strategies, intended to facilitate the development of renewable energy–technology manufacturing and service industries in the host jurisdictions. Such goals were central to the rationale for Ontario's 2009 *Green Energy and Green Economy Act* (GEA) (Weis et al. 2011). The role of the GEA as an industrial strategy has also been a central issue in critiques of the initiative (McKitrick 2013).

This chapter examines the debates over the GEA as an industrial strategy through the lenses of the different ideational perspectives that have defined the discourse over the legislation. The chapter then assesses the province's approach to the development of the renewable energy–technology sector in the aftermath of the act's adoption.

ANALYTICAL APPROACH: THE ROLE OF IDEATIONAL PERSPECTIVES IN GREEN ENERGY DEBATES

An important feature of the debates surrounding renewable energy initiatives is that they are not necessarily limited to questions directly related to energy policy. Rather, they are embedded in wider ideological debates about the appropriate roles of government, public policy, and markets in achieving societal goals. The role of ideational perspectives is particularly important in the case of the debates over the GEA. The available empirical data on the economic and employment impacts of the initiative is very limited. Instead, the evidence regarding the GEA's economic impacts is almost entirely grounded in economic modelling exercises, exercises that may reflect the ideological perspectives of the modellers (Winfield et al. 2013).

Several different ideational views figure prominently in the debates over the economic development impact of renewable energy initiatives. These include market fundamentalism, economic rationalism, ecological modernism, and progressive political economy perspectives (Dryzek 2013; Winfield and Dolter 2014). Market fundamentalists, as represented by various non-governmental think-tanks, have been among the most prominent public critics of renewable energy initiatives. These actors tend to be ideologically opposed to any form of governmental intervention in the market and have found renewable energy initiatives particularly objectionable in this context. Economic rationalists are generally committed to the intelligent use of market mechanisms to achieve public ends and are often neo-classically grounded academic economists. They have also been important critics of renewable energy initiatives, arguing that they are an inefficient means of achieving environmental and economic policy goals, although they are not necessarily ideologically opposed to interventions in markets for these purposes.

Ecological modernists, on the other hand, generally favour a restructuring of the capitalist political economy in a more environmentally sustainable direction and an active role for the state in those processes. They have tended to support renewable energy initiatives as expressions of the movement in precisely this direction. Although the concept of ecological modernism is less well developed in Canada than in Western Europe, it does overlap with the progressive political economy stream of Canadian academic and labour economists. Individuals and organizations in the latter camp tend to argue for public policies that enhance the development of high-value-added, innovative industrial sectors in Canada (Stanford 2012). A wider resurgence of interest in industrial policy in Canada and elsewhere in the Organisation for Economic Co-operation and Development has also been noted recently (Ciuriak and Curtis 2013).

ONTARIO'S *GREEN ENERGY AND GREEN ECONOMY ACT*

The GEA was adopted in May 2009 under the leadership of then minister of energy George Smitherman. The centrepiece of the GEA initiative was the FIT program established under the legislation, which provided stable prices under long-term contracts for energy generated from renewable sources – specifically solar, wind, biomass, biogas, and water power. The Ontario Power Authority (OPA) was given responsibility for implementing the FIT program, entering into contracts with eligible applicants. The program was divided into two categories, FIT and MicroFIT, with the FIT program intended for projects over 10 kilowatts (kW) and the MicroFIT program for projects less than 10 kW. Some of the key design features of the FIT program are outlined in Table 5.1 below.

The province terminated the FIT program for projects over 500 kW in May 2013 in favour of competitive bidding processes, in part due to concerns over program costs (Ontario Ministry of Energy 2013b).

Domestic Content Requirements

All FIT projects were initially required to include a minimum amount of goods and services made in Ontario, and following the Ministry of Energy's review of the FIT program in 2012, the domestic content requirements were 50 percent for solar projects and 60 percent for wind projects (OPA 2013). However, as a result of the World Trade Organization (WTO) ruling regarding the Ontario FIT domestic content requirements on 24 May 2013, the requirements were reduced to between 19 and 28 percent, depending on the wind and solar photovoltaics (PV) technologies involved (Chiarelli 2013), and subsequently removed completely (Spears 2013a).

TABLE 5.1
FIT Rates

Renewable fuel	*Project size tranche*	*Original (2009) FIT price (cents per kWh)*	*FIT price (cents per kWh), 5 April 2012*	*FIT price (cents per kWh), 26 August 2013*
Solar (PV) rooftop	≤ 10 kW	80.2	54.9	39.6
	> 10 ≤ 100 kW	71.3	54.8	34.5
	> 100 ≤ 500 kW	63.5	53.9	32.9
	> 500 kW	53.9	48.7	n/a
Solar (PV) non-rooftop	≤ 10 kW	64.2	44.5	29.1
	> 10 ≤ 500 kW	44.3	38.8	28.8
	> 500 kW ≤ 5 MW	44.3	35.0	n/a
	> 5 MW	44.3	34.7	n/a
Onshore wind	All sizes	13.5	11.5	11.5
Water power	≤ 10 MW	13.1	13.1	14.8
	> 10 MW ≤ 50 MW	12.2	12.2	14.8
Renewable biomass	≤ 10 MW	13.8	13.8	15.6
	> 10 MW	13.0	13.0	15.6
On-farm biogas	< 100 kW	19.5	19.5	26.5
	100 ≤ 250 kW	18.5	18.5	21.0
Biogas	≤ 500 kW	16.0	16.0	16.4
	> 500 kW ≤ 10 MW	14.7	14.7	16.4
	> 10 MW	10.4	10.4	16.4
Landfill gas	≤ 10 MW	11.1	11.1	7.7
	> 10 MW	10.3	10.3	7.7

Source: Author's compilation from OPA (2014).

Incentive for Aboriginal and Community Groups

As a result of the 2012 FIT review, security payments were decreased for Aboriginal- and community-owned projects. The program also included incentives for projects with significant Aboriginal or community participation: 0.75 to 1.5 cents per kWh for projects with Aboriginal participation and 0.5 to 1 cent per kWh for projects with community participation.

Streamlined Regulatory Approvals Process

In 2012, the province established a Renewable Energy Approval process. It provided for consolidated environmental approvals of renewable energy projects and exempted FIT-supported projects from municipal planning-approval requirements (Mulvihill, Winfield, and Etcheverry 2013).

The FIT program was to function within the targets and parameters set out in the province's 2010 Long-Term Energy Plan (LTEP). According to the 2010 LTEP, 50 percent of Ontario's demand was to be met by nuclear power and 13 percent by wind, solar, and bioenergy by 2018 (Ontario Ministry of Energy and Infrastructure 2010). The most recent Supply Mix Directive (February 2011) from the minister of energy specified a target of 10,700 megawatts (MW) of renewable generation, excluding hydroelectric, by 2018 (Duguid 2011).

ECOLOGICAL MODERNISM AND THE EMERGENCE OF GREEN INDUSTRIAL DEVELOPMENT STRATEGIES

It is important to place the debates over the economic role of renewable energy technologies within the wider context of the evolution of the conceptualization of the place of environmental services and technologies in the economy. Environmental technologies and services were initially conceived of as facilitative adjuncts to economic growth and development, with the latter understood in conventional terms: industrialization, urbanization, and resource extraction and processing. The focus was on add-on, end-of-pipe pollution control technologies, intended to mitigate the worst and most obvious environmental impacts of industrial activities in order to render them more socially acceptable. Investments in environmental technologies were seen as regrettable but necessary costs of doing business from the viewpoints of governments and industrial operators. Government policy gave little or no formal recognition to environmental or green technologies and services as a distinct sector of economic activity.

This view dominated in North America from the time that environmental pollution was initially recognized as a significant public and health concern, in the second half of the 19th century, until the late 1970s. Views on the role of environmental or green technologies began to shift from that point on. In North America, the efforts of the Canada-US International Joint Commission highlighted the long-term ineffectiveness of end-of-pipe pollution control technologies, while at the global level, the work of the World Commission on Environment and Development (the Brundtland Commission) emphasized economic and environmental interdependence through the concept of sustainable development (World Commission on Environment and Development 1987).

Modifying industrial activities to prevent the generation of pollutants and improve their energy, materials, and water efficiency was seen as offering the potential both to reduce extractive and assimilative pressures on local environments and the global biosphere and to increase the productivity of economic activities (MacNeill, Winsemius, and Yakushiji

1991). Northern European countries, whose energy and material security constraints had been highlighted during the energy shocks of the 1970s, generally recognized these potential connections earlier than was the case in North America (Hay 2005). Ontario formally recognized the potential positive linkages between economic and environmental policies in the early 1990s. It also identified environmental services and technologies as a distinct sector of the economy and created a first generation of strategies for its development (Ontario Green Industry Ministerial Advisory Committee 1994).

The most recent stage in the evolution of the role of environmental, or green, industries has shifted these activities from being adjuncts to the mainstream economy to a much more central position. While the focus on pollution prevention and the energy, water, and materials efficiency of conventional economic activities continues, increasing attention is being paid to the potential role of the design, development, manufacturing, installation, and servicing of green technologies, particularly renewable energy technologies, as major components of the industrial economy in themselves. This ecological modernist vision is grounded in the apparent success of countries like Finland, Norway, Sweden, Denmark, Germany, and the Netherlands. These countries have combined the retention of substantial value-added manufacturing activities, in which environmental technologies figure significantly, with consistently high rankings in measures of environmental performance (Dryzek 2013).

In Denmark, for example, employment in the wind sector approaches 30,000 individuals (2009 figure), principally in design, manufacturing, and service-based activities (Danish Wind Industry Association 2010), while the energy technology sector accounts for 11 percent of the country's total manufacturing economy (CEPOS 2009). Employment in Germany's renewable energy sector in 2011 was placed at 381,600, again strongly weighted in the direction of value-added design and manufacturing activities.[1] The sector experienced substantial growth in employment in both countries over the second half of the 1990s.

ECOLOGICAL MODERNISM COMES TO NORTH AMERICA

The apparent success of these jurisdictions had a major influence in North America on the formulation of policy responses to the challenges of mitigating climate change and the impact of the 2008 economic downturn on manufacturing activities. In announcing the energy and climate change leaders for his incoming administration, US president-elect Barack Obama made the following statement about the potential for his administration to integrate energy, environmental, and economic objectives:

> We can seize boundless opportunities for our people. We can create millions of jobs, starting with a 21st-century economic recovery plan that puts

> Americans to work building wind farms, solar panels, and fuel-efficient cars.
>
> We can spark the dynamism of our economy through a long-term investment in renewable energy that will give life to new businesses and industries with good jobs that pay well and can't be outsourced.
>
> We'll make public buildings more efficient, modernize our electricity grid, and reduce greenhouse gas emissions, while protecting and preserving our natural resources. (Obama 2008)

The pursuit of similar objectives was a major rationale for Ontario's GEA. In particular, the relatively generous rates built into the original FIT program established under the legislation were intended to facilitate the rapid development of a critical mass of activity in the province's renewable energy sector (Amin 2012). The provincial government and supporters of the legislation hoped that the strong domestic market produced by the FIT program would provide the foundation for a renewable energy–technology manufacturing and services sector that would then be able to sell its products and services beyond the province's borders.

At the time the GEA was being developed, a number of factors suggested that such a strategy could be successful despite the relative dominance of European, particularly Danish and German, suppliers in the international renewable energy–technology and services market. Ontario had the advantage of relative proximity to and a long-established relationship with the US market, where the incoming federal administration, as noted earlier, was signalling its intention to make major investments in the development of renewable energy sources (Weis and Bramley 2009). Many US state governments were also indicating their interest in the rapid and large-scale development of renewable energy resources (Rabe 2010a, 2010b). At the same time, wind turbine prices were rising substantially as global and North American demand began to outstrip the existing capacity of the established manufacturers (Wiser and Bolinger 2012). Ontario's historical strengths in mechanical and electro-mechanical engineering, design, and manufacturing – products of the province's long-standing engagement with the production of transportation equipment – were seen as being potentially transferable to producing renewable energy technology, particularly wind turbines (*Globe and Mail* 2010). These strengths could provide the province with a comparative advantage relative to the US states that were contemplating moves into the renewable energy–technology supply and services sector themselves. At the same time, the interest of these states reinforced the need for Ontario to establish a presence in the sector relatively quickly.

THE IDEOLOGICAL DEBATE: GREEN ENERGY, MARKET FUNDAMENTALISTS, PROGRESSIVE POLITICAL ECONOMISTS, AND INDUSTRIAL STRATEGY

The debates concerning the industrial development rationale for renewable energy initiatives like the GEA are embedded in wider discourses about the appropriate role of government in the development of specific industries and sectors. The most prominent and vociferous public critics of this aspect of renewable energy initiatives tend to represent the market fundamentalist school of thought. These critics take the view that such strategies are almost certain to be unsuccessful, grounded in a belief that government is much less efficient and effective than the market at picking potential economic winners and losers. The author of a recent Fraser Institute critique of the GEA noted, for example, in an article in the *Financial Post,*

> With regards to job creation, there is nothing special about subsidizing electricity generation. It's just as harmful as subsidizing anything else. We have long and lamentable experience in Canada with failed job creation schemes based on subsidies to money-losing industries. From Sprung cucumbers to Bricklin sports cars, governments have regularly learned and relearned, at taxpayer expense, the immutable rule that if a business plan depends on subsidies, the jobs it creates are not sustainable, and if the business is profitable on its own, it doesn't need subsidies. (McKitrick 2011)

On the other hand, those more sympathetic to renewable energy initiatives tend to take the view that advanced industrial economies need to pursue active industrial strategies to retain and build high-value-added economic activities. Researchers in this progressive political economy camp highlight the presence of active industrial strategies in the northern European economies (e.g., Germany, Denmark, Sweden, Finland), which have retained significant manufacturing activities and a role for green technologies in that process (Dryzek 2013; O'Sullivan et al. 2012, Stanford 2012). The development of green skills and jobs has emerged as a significant sub-theme within this school of thought (Lee and Card 2012).

THE GEA AS A SECTORAL DEVELOPMENT STRATEGY

The potential for the development of a renewable energy–technology manufacturing and services industry in Ontario was a fundamental rationale for adopting the GEA. However, the experiences of European jurisdictions that have succeeded in developing substantial renewable energy manufacturing lend some important considerations to the design of such a strategy in Ontario. Studies of the German and Danish renewable energy industries highlight the need to move beyond the domestic markets, whose emergence was spurred by FIT programs, in order to foster a viable upstream renewable energy–technology industry. In the long term, export markets are consistently identified as the key source of employment growth in the renewable energy sector in these jurisdictions (Danish Wind Industry Association 2010; Lehr et al. 2012).

The implication for Ontario is that the FIT program alone, whose primary impact would be developing a domestic market for renewable energy technologies, would not be sufficient to sustain a renewable energy manufacturing and services sector in the province. Rather, the development of a domestic market would need to be complemented by an active sectoral development strategy to identify and develop markets outside of Ontario.

The government of Ontario has considerable experience in developing sectoral strategies, a concept first introduced in the early 1990s under the New Democratic Party government of then premier Bob Rae (Rachlis and Wolfe 1997; Ontario Ministry of Finance 1991; Ontario Ministry of Industry, Trade and Technology 1992). The original sectoral strategies were typically structured around sectoral councils, with representation from the sector and related industry interests as well as from labour, non-governmental organizations, and academia. The councils were given research and institutional support through the relevant provincial government agencies, and they were mandated to develop strategies for developing their sectors, focusing on measures that the provincial government could take to support those efforts.

Although widely regarded as one of the most effective initiatives of the Rae government (Courchene and Telmer 1998), the concept of sectoral strategies was abandoned during the Progressive Conservative governments of Mike Harris and Ernie Eves in favour of a simplified approach focused on tax cuts and removing regulatory burdens on industry. The sectoral approach re-emerged under Liberal premier Dalton McGuinty (Winfield 2012). Recent strategies related to the mining (OMIC n.d.) and financial services sectors (Ontario Open for Business 2013) have been highlighted as being particularly successful (Radwanski, Kiladze, and Perkins 2012).

Unfortunately, no such sectoral development strategy accompanied the GEA when it was adopted in 2009. Rather, there was a series of relatively ad hoc efforts to promote the development of the upstream manufacturing and services components of the sector. A January 2010 agreement with the South Korean industrial giant Samsung exchanged guarantees of a portion of available FIT contracts for promises of investment in renewable energy–technology manufacturing activities in Ontario (Canwest News Service 2010).[2] Domestic content requirements were also incorporated into the original FIT program to promote the development of a renewable energy industry in the province – requiring, as noted earlier, that a minimum portion of the capital costs of FIT-contracted projects be sourced in Ontario. The domestic content requirements were subsequently subject to a successful challenge under WTO rules by the EU, Japan, and the United States (McCarthy 2013). Despite the scale of the investments being directed toward the sector, the lack of any apparent overall strategy to develop the renewable energy sector beyond these measures prompted the following observation:

> The Ontario government touts its intention to become a leader in exporting clean energy technologies, portraying these technologies as one of the province's strengths. However, its current policy framework is not designed to support this aim. (Khanberg and Joshi 2012, 44)

In fact, formal consideration of a sectoral strategy for the renewable energy sector in Ontario did not occur until the two-year review of the FIT program, initiated in October 2011. The review report, authored by the deputy minister of energy, concluded that the program should continue, and potentially be expanded, subject to reductions in the rates paid for some types of FIT projects and a strengthening of mechanisms that favoured projects initiated or supported at the community level (Amin 2012). The report's key recommendation from an economic development perspective was to propose the development of a Clean Energy Economic Development Strategy – effectively a sectoral development strategy for the renewable energy sector. The report specifically recommended that the province do the following:

- Provide targeted financial support through the Smart Grid Fund to Ontario-based demonstration and capacity-building projects that test, develop, and bring to market the next generation of technology solutions.
- Work with key stakeholders to consider the potential for a clean energy institute to spur domestic innovation and achieve a greater global market presence for Ontario-based companies.
- Support domestic manufacturers by showcasing Ontario's smart energy solutions through a strategic export strategy.
- Create a Clean Energy Task Force to advise the ministers of energy and economic development and innovation on potential strategies for Ontario's clean energy sector.

The establishment of the Clean Energy Task Force and strategy was announced the following month. The task force was mandated to "help broaden Ontario's energy focus by facilitating collaboration within Ontario's clean energy industry to identify export markets, marketing opportunities and approaches to demonstrate Ontario's advanced clean energy systems" (Ontario 2012). The province also committed to leading clean-technology trade missions that would support domestic manufacturers by showcasing Ontario's clean energy solutions in key markets, including Asia, the Middle East, and the United States, and to delivering on its Smart Grid Fund and other targeted investments to spur innovation in priority areas (ibid.).

ANALYSIS AND DISCUSSION

The debates around the notion of the GEA as an industrial development strategy are grounded in wider ideological arguments about the appropriate roles of markets and governments in economic policy. The most aggressive public critics of the GEA have tended to come from a strong market fundamentalist orientation, and they have a record of opposing any form of government intervention in the marketplace (e.g., the Fraser Institute). Supporters of the initiative, on the other hand, reflect ecological modernist and progressive political economy perspectives. These schools of thought highlight the importance of government interventions in the economy to counteract the pull of dependence on commodity exports in relatively resource-rich economies like Canada's and to support the development of a more diversified economy grounded in the provision of value-added goods and services.

At the time the GEA initiative was formulated, there was considerable potential for major growth in Ontario and US demand for renewable energy technologies. Local and global shortfalls in supply and manufacturing capacity for these technologies were emerging at the same time. In combination with the province's historical strength in related engineering and manufacturing activities, there was a potential for the province to establish itself as a significant player in the sector. However, even at that stage, the challenges to the successful pursuit of such a strategy were considerable. Other jurisdictions pursuing the development of renewable energy resources were likely to prefer domestically sourced equipment wherever possible. More recently, the entry of China into the renewable energy–technology supply market is posing challenges even for long-established players like Germany and Denmark (Cao and Groba 2013). The three-year delay between adopting the GEA and the beginning of the establishment of a coherent strategy for developing the renewable energy–technology and services sector is likely to have cost Ontario important opportunities.

Even with the establishment of the Clean Energy Economic Development Strategy, the upstream renewable energy industry in Ontario continues to face significant domestic challenges as well. Increasing uncertainty over the province's direction with respect to renewable energy has worked to discourage long-term investment in manufacturing capacity. Changes in program rules and delays in processing project applications and obtaining connections to the grid have all reinforced these concerns. Political conflicts over the impact of the FIT program, which raised questions about its continued existence beyond the October 2011 election, and the more than year-long moratorium on FIT applications while the program was under review following that election added to the doubts about the program's future (Hamilton 2012; Holburn, Lui, and Morand 2009; Strifler 2012).

At the same time, the build-out of the renewable energy capacity contracted between 2009 and 2011 remains to be completed.[3] However, once this is accomplished (by 2018, if not earlier), there is no certainty regarding the province's longer-term intentions with respect to renewable energy. In fact, the province's most recent (2013) LTEP, which outlines the province's plans to 2030, implies that all growth in renewable energy generation will be completed by 2018 (Ontario Ministry of Energy 2013a). Questions about the potential for future growth in renewables are reinforced by projections of little or no growth in electricity demand for the foreseeable future (IESO 2014a) and the government's continued commitment to nuclear energy, particularly the refurbishment of the Darlington and Bruce nuclear facilities. The lack of any fully developed strategies for the role of smart-grid technologies in the integration of intermittent renewable energy technologies into the electricity system further complicates the picture (Winfield and Weiler 2014), as does the absence, until very recently, of strategies for the development of grid-scale energy storage resources (Energy Storage Ontario 2014). Without such strategies, the usefulness of the renewable energy resources that are under development in the province may be limited (SEI 2012).

The government's May 2013 announcement that it was terminating the FIT program for all but small projects (under 500 kW) in favour of competitive bidding processes reinforces those concerns, as does its commitment to allocate the remaining 900 MW of capacity space available until 2018 to smaller renewable energy projects with municipal participation, with no indication of any commitments to renewables beyond that date (Ontario Ministry of Energy 2013b). The announcement a few weeks later that the province was dropping its commitment to purchase renewable energy with Samsung from 2,500 MW to 1,369 MW has had the same effect. With a 45 percent reduction relative to the original January 2010 agreement, it is unclear whether the difference in energy supply will be obtained from other sources or removed from the province's energy plans completely (Spears 2013b). A Danish-style outcome may still be possible, whereby a strong export industry is built on the basis of rapid domestic development, but the province's domestic market is now relatively weak. The window for pursuing a similar strategy is now very narrow and would require a very well-developed strategy for a still relatively nascent sector.

THE NEXT STEPS FOR ONTARIO'S GREEN ENERGY STRATEGY

A better and less risk-laden outcome would be for the province to clarify its commitment to renewable energy development beyond the 2018 target specified in the LTEP. More broadly, future efforts at system planning need to proceed on a level-playing-field basis, considering the externalities, risks, liabilities, and potential contributions to the sustainability associated with all available technologies on a life-cycle basis. At the same time, if Ontario intends to continue to pursue the development of the renewable energy sector, it needs to advance its clean-energy economic development strategy to take advantage of the 2013–18 build-out of existing contracts. Among other things, it will need to do the following:

- Develop a comprehensive, empirically based profile of the renewable–energy technology and services sector in Ontario, similar to the profiles developed by other jurisdictions pursuing the development of their renewable energy sectors.
- Identify areas of potential comparative advantage in renewable energy technology and services for Ontario.
- Assess potential external markets for the Ontario industry in Canada, the United States, and overseas; this will include closely monitoring policy and program commitments as well as supply chains in those markets.
- Assess the education and skills development requirements in the sector and develop appropriate mechanisms to ensure that Ontario's post-secondary institutions address these needs.
- Support market development as well as research and development, as outlined in the deputy minister's 2012 FIT review report.
- Develop and implement energy-storage and smart-grid strategies to support the integration of renewable energy resources into the province's energy systems up to their full potential.

If these steps are not taken, the province runs the considerable risk that, from an economic development perspective, the GEA exercise will amount to an expensive but temporary counter-cyclical intervention as opposed to an investment in developing an industrial sector with the potential to make significant long-term contributions to the sustainability of Ontario's economy and environment.

NOTES

1. The sector employed 74,000 people in manufacturing and production of onshore wind technologies versus 17,800 in operation and maintenance; and 103,000 workers in solar PV production versus 7,600 in operation and maintenance (O'Sullivan et al. 2012).
2. Under the agreement, 2,500 MW of renewable energy capacity was dedicated to Samsung in exchange for it building four manufacturing plants, which would create 1,440 jobs. The government claimed that the agreement as a whole would create approximately 16,000 jobs.

3. The March 2012 FIT review report gives the figure of 7,100 MW contracted through the FIT program – 4,600 MW through the FIT itself and 2,500 MW through the January 2010 Samsung agreement (Amin 2012). In May 2013, the Independent Electricity System Operator (IESO) reported that 1,560 MW of wind capacity had been installed in Ontario and that installed capacity of solar and biomass amounted to 122 MW (IESO 2014b). The June 2013 modifications to the original Samsung agreement reduced the amount of committed renewable energy by 1,131 MW, and it is unclear whether this will be reallocated to other suppliers or removed from the plan.

REFERENCES

Amin, F. 2012. *Ontario's Feed-in Tariff Program: Two-Year Review Report*. Toronto: Ministry of Energy. http://www.energy.gov.on.ca/en/files/2014/09/FIT-Review-Report-en.pdf.

Canwest News Service. 2010. "Ontario Signs Green Energy Deal with Samsung Team." *Financial Post*, 21 January. http://www.financialpost.com/story.html?id=2468582.

Cao, J., and F. Groba. 2013. *Chinese Renewable Energy Technology Exports: The Role of Policy, Innovation and Markets*. Discussion Paper 1263. Berlin: Deutsches Institut für Wirtschaftsforschung [German Institute for Economic Research]. http://www.diw.de/documents/publikationen/73/diw_01.c.414422.de/dp1263.pdf.

CEPOS (Center for Politiske Studier). 2009. *Wind Energy: The Case of Denmark*. Copenhagen: CEPOS.

Chiarelli, B. 2013. "RE: Administrative Matters Related to Renewable Energy and Conservation Programs." Letter sent to Colin Andersen, chief executive officer, OPA. 16 August. http://www.powerauthority.on.ca/sites/default/files/page/DirectionAdministrativeMatters-renewables-Aug16-2013.pdf.

Ciuriak, D., and J.M. Curtis. 2013. "The Resurgence of Industrial Policy and What It Means for Canada." *IRPP Insight* 2 (June).

Courchene, T., and C.R. Telmer. 1998. *From Heartland to North American Regional State: The Social, Fiscal and Federal Evolution of Ontario*. Toronto: University of Toronto Press.

Danish Wind Industry Association. 2010. "Danish Wind Industry: Annual Statistics 2010." http://ipaper.ipapercms.dk/Windpower/Branchestatistik/DanishWindIndustryAnnualStatistics2010/.

Dryzek, J. 2013. *The Politics of the Earth: Environmental Discourses*. Oxford: Oxford University Press.

Duguid, B. 2011. "Supply Mix Directive." Sent to Colin Andersen, chief executive officer, OPA. 17 February. http://www.powerauthority.on.ca/sites/default/files/new_files/IPSP%20directive%2020110217.pdf.

Edge, G. 2006. "A Harsh Environment: The Non-fossil Fuel Obligation and the UK Renewables Industry." In *Renewable Energy Policy and Politics: A Handbook for Decision-Making*, ed. K. Mallon, 163–84. London: Earthscan.

Energy Storage Ontario. 2014. "Ontario Electricity Market." http://www.energystorageontario.com/ontario-market/.

Globe and Mail. 2010. "Linamar Partners on Wind Turbines." 5 May. http://www.theglobeandmail.com/globe-investor/linamar-partners-on-wind-turbines/article1390170/.

Hamilton, T. 2012. "Ontario Teaches World How Not to Run a FIT Program." *Thestar.com*, 5 October. http://www.thestar.com/business/2012/10/05/ontario_teaches_world_how_not_to_run_a_fit_program.html.

Hay, C. 2005. "EU Environmental Policies: A Short History of the Policy Strategies." In *EU Environmental Policy Handbook*, ed. S. Scheuer, 17–30. Brussels: European Environment Bureau.

Holburn, G., K. Lui, and C. Morand. 2009. "Policy Risk and Private Investment in Ontario's Wind Power Sector." Policy paper. London: Ivey School of Business, University of Western Ontario. http://sites.ivey.ca/energy/files/2010/04/Holburn-Lui-Morand-Ontario-Wind-Policy-Canadian-Public-Policy.pdf.

IESO (Independent Electricity System Operator). 2014a. "18-Month Outlook: From December 2014 to May 2016." https://ieso-public.sharepoint.com/Documents/marketReports/18Month_ODF_2014nov.pdf.

—. 2014b. "Supply Overview." http://ieso-public.sharepoint.com/Pages/Power-Data/Supply.aspx.

IPCC (Intergovernmental Panel on Climate Change). 2011. *Renewable Energy Sources and Climate Change Mitigation: Special Report of the Intergovernmental Panel on Climate Change*. Cambridge: Cambridge University Press.

Jacob, D. 2012. *Renewable Energy Policy Convergence in the EU*. Burlington, VT: Ashgate.

Khanberg, T., and R. Joshi. 2012. *Smarter and Stronger: Taking Charge of Canada's Energy Technology Future*. Toronto: Mowat Centre, School of Public Policy and Governance, University of Toronto.

Lee, M., and A. Card. 2012. *A Green Industrial Revolution: Climate Justice, Green Jobs and Sustainable Production in Canada*. Ottawa: Canadian Centre for Policy Alternatives.

Lehr, U., B. Breitschopf, J. Diekmann, J. Horst, M. Klobasa, F. Sensfuss, and J. Steinbach. 2012. *Renewable Energy Deployment: Do the Benefits Outweigh the Costs?* GWS Discussion Paper 2012/5. Osnabrück: Gesellschaft für Wirtschaftliche Strukturforschung [Institute for Economic Structures Research].

MacNeill, J., P. Winsemius, and T. Yakushiji. 1991. *Beyond Interdependence: The Mashing of the World's Economy and the Earth's Ecology*. New York: Oxford University Press.

McCarthy, S. 2013. "Ontario Loses Final WTO Appeal on Green Energy Act." *Globe and Mail*, 6 May, B3.

McKitrick, R. 2011. "Ontario's Power Trip: The Failure of the Green Energy Act." *Financial Post*, 16 May. http://opinion.financialpost.com/2011/05/16/ontarios-power-trip-the-failure-of-the-green-energy-act/.

—. 2013. *Environmental and Economic Consequences of Ontario's Green Energy Act*. Ontario Prosperity Initiative. Vancouver: Fraser Institute. http://www.fraserinstitute.org/uploadedFiles/fraser-ca/Content/research-news/research/publications/environmental-and-economic-consequences-ontarios-green-energy-act.pdf.

Mulvihill, P., M. Winfield, and J. Etcheverry. 2013. "Strategic Environmental Assessment and Advanced Renewable Energy in Ontario: Moving Forward or Blowing in the Wind." *Journal of Environmental Assessment Policy and Management* 15 (2):1.

Obama, B. 2008. "Obama's Energy and Environment Team Announcement." *New York Times*, 16 December. http://www.nytimes.com/2008/12/16/world/americas/16iht-15textobama.18710816.html?pagewanted=all&_r=0.

OMIC (Ontario Mineral Industry Cluster). n.d. "Ontario's Mineral Industry Cluster: An Economic Powerhouse." Brochure. Toronto: OMIC.

Ontario. 2012. "Expanding Ontario's Clean Energy Economy: McGuinty Government Launches Clean Energy Economic Development Strategy." News release, 12 April. http://news.ontario.ca/mei/en/2012/04/expanding-ontarios-clean-energy-economy.html.

Ontario. Green Industry Ministerial Advisory Committee. 1994. "Ontario's Green Industry Strategy for a Clean Environment & Strong Economy." Toronto: Ministry of Energy and the Environment.

Ontario. Ministry of Energy. 2013a. *Achieving Balance: Ontario's Long-Term Energy Plan*. Toronto: Queen's Printer. http://www.energy.gov.on.ca/en/ltep/.

—. 2013b. "Ontario Working with Communities to Secure Clean Energy Future." News release, 30 May. http://news.ontario.ca/mei/en/2013/05/ontario-working-with-communities-to-secure-clean-energy-future.html.

Ontario. Ministry of Energy and Infrastructure. 2010. *Achieving Balance: Ontario's Long-Term Energy Plan*. Toronto: Ministry of Energy and Infrastructure. http://www.energy.gov.on.ca/en/ltep/.

Ontario. Ministry of Finance. 1991. *1991 Budget*. Ministry of Finance Special Publication, Budget Paper E. Toronto: Queen's Printer for Ontario.

Ontario. Ministry of Industry, Trade and Technology. 1992. *An Industrial Policy Framework for Ontario*. Toronto: Queen's Printer for Ontario.

Ontario. Open for Business. With Toronto Financial Services Alliance. 2013. "Ontario's Business Sector Strategy: Financial Services Sector." Toronto: Queen's Printer for Ontario.

OPA (Ontario Power Authority). 2013. "Domestic Content." http://fit.powerauthority.on.ca/program-resources/faqs/domestic-content.

—. 2014. "FIT Price Schedule." http://fit.powerauthority.on.ca/fit-program/fit-program-pricing/fit-price-schedule.

O'Sullivan, M., D. Edler, T. Nieder, T. Rüther, U. Lehr, and F. Peter. 2012. *Employment from Renewable Energy in Germany: Expansion and Operation – Now and in the Future*. First Report on Gross Employment. Bonn: Federal Ministry for the Environment, Nature Conservation and Nuclear Safety. http://www.erneuerbare-energien.de/fileadmin/ee-import/files/english/pdf/application/pdf/ee_bruttobeschaeftigung_en_bf.pdf.

Rabe, B. 2010a. "The Aversion to Direct Cost Imposition: Selecting Climate Policy Tools in the United States." *Governance: An International Journal of Policy, Administration, and Institutions* 23:583.

—. 2010b. *Greenhouse Governance: Addressing Climate Change in America*. Washington, DC: Brookings Institution Press.

Rachlis, C., and D. Wolfe. 1997. "An Insider's View of the NDP Government of Ontario: The Politics of Permanent Opposition Meets the Economics of Permanent Recession." In *The Government and Politics of Ontario*, 5th ed., ed. G. White, 348–51. Toronto: University of Toronto Press.

Radwanski, A., T. Kiladze, and T. Perkins. 2012. "What's Wrong with Ontario – and How to Make It Right." *Globe and Mail*, 18 February. http://www.theglobeandmail.com/report-on-business/economy/whats-wrong-with-ontario---and-how-to-make-it-right/article550134/?page=all.

SEI (Sustainable Energy Initiative). 2012. "Storage Options for Renewable Energy: Developing Renewable Energy to Commercialization." Video, 2:01:46. 21 September. http://www.youtube.com/watch?v=vx4gXvUTZ90.

Spears, J. 2013a. "Ontario Liberals Scrap Local Content Rules for Green Energy." *Toronto Star*, 12 December.

—. 2013b. "Samsung Green Deal Scaled Back." *Toronto Star*, 21 June, B1.

Stanford, J. 2012. "A Cure for Dutch Disease: Active Sector Strategies for Canada's Economy." Alternative Federal Budget 2012. Technical Paper. Ottawa: Canadian Centre for Policy Alternatives.

Strifler, D. 2012. "Small Scale, Big Impact: A Comprehensive Evaluation of Ontario's MicroFIT Program." Unpublished research paper, Faculty of Environmental Studies, York University, Toronto.

US (United States). Energy Information Administration. 2012. "Most States Have Renewable Portfolio Standards." 3 February. http://www.eia.gov/todayinenergy/detail.cfm?id=4850.

Weis, T., and M. Bramley. 2009. "United States to Invest Over Six Times More per Capita in Renewable Energy and Energy Efficiency Than Canada." Media backgrounder. Gatineau: Pembina Institute. http://www.pembina.org/reports/canada-v-us-investment-in-re-ee-backgrounder.pdf.

Weis, T., P. Gipe, Shine ONtario, and Green Energy Act Alliance. 2011. *Ontario Feed-in Tariff: 2011 Review*. FIT Review Joint Submission. Toronto: Pembina Institute. http://www.pembina.org/reports/on-feed-in-tarif-2011-review.pdf.

Winfield, M. 2012. *Blue-Green Province: The Environment and the Political Economy of Ontario*. Vancouver: University of British Columbia Press.

Winfield, M., and B. Dolter. 2014. "Energy, Economic and Environmental Discourses and Their Policy Impact: The Case of Ontario's Green Energy and Green Economy Act." *Energy Policy* 68:423–35.

Winfield, M., R. Gibson, T. Markvart, K. Gaudreau, and J. Taylor. 2010. "Implication of Sustainability Assessment for Electricity System Design: The Case of the Ontario Power Authority's Integrated Power System Plan." *Energy Policy* 38:4115–26.

Winfield, M., N. Rehman, M. Eret, D. Strifler, and P. Cockburn. 2013. *Understanding the Economic Impact of Renewable Energy Initiatives: Assessing Ontario's Experience in a Comparative Context*. Studies in Ontario Energy Policy Series. Toronto: Sustainable Energy Initiative, Faculty of Environmental Studies, York University.

Winfield, M., and S. Weiler. 2014. "Beyond Smart Meters: The State of Ontario Smart Grid Policy and Practice." Working paper. Toronto: Sustainable Energy Initiative, Faculty of Environmental Studies, York University. http://sei.info.yorku.ca/files/2012/12/Beyond-Smart-Meters-April-28-2014.pdf.

Wiser, R., and M. Bolinger. 2012. *2011 Wind Technologies Market Report*. Washington, DC: US Department of Energy.

World Commission on Environment and Development. 1987. *Our Common Future*. Toronto: Oxford University Press.

II

MAKING GREEN WORK

CHAPTER 6

(RE)BUILDING SUSTAINABLE INFRASTRUCTURE: THE IMPLICATIONS FOR ENGINEERS

KEAN BIRCH AND DALTON WUDRICH

INTRODUCTION

From Hurricane Sandy in 2012 to the Calgary floods in 2013, several major weather- or climate-related events have made the news headlines around the world in the last year or so. Such events are newsworthy not only because of the deaths they have caused but also because they entailed significant damage to the physical infrastructure that underpins our economies and communities. Thus, it is becoming evident that changing weather and climate will have an increasingly dramatic impact on our lives over the next few decades as a result of changing demands on and damage to our infrastructure systems (Deveau 2012; Wells 2012; Justian 2013; McLaughlin 2013). This means that we need to think about and prepare for these impacts now rather than at some unspecified time down the road.

Broadly speaking, this chapter is concerned with the implications of climate change for core infrastructure systems – including transit, roads, energy, and water – and for the engineering profession that plans, designs, builds, and maintains them. In the 2006 Stern Review, Lord Nicholas Stern, professor at the London School of Economics, argued that we need to take "strong action now … to avoid the worst impacts of climate change" (Stern 2006, vi). Part of this action involves rethinking and rebuilding our infrastructure systems so that they are sustainable,

Work in a Warming World, edited by Carla Lipsig-Mummé and Stephen McBride. Kingston: School of Policy Studies, Queen's University.

adaptable, and resilient. Engineers do and will play an important role in this transformation, and it is critical to understand what implications this role has for the engineering profession.

What makes this situation particularly urgent to consider now is that, by 2030, Canada and other countries around the world will have to spend trillions of dollars to renew or build new infrastructure (Infrastructure Canada 2006; FCM 2007; Engineers Canada 2008; Corfee-Morlot et al. 2012). This renewed and new infrastructure will have to be both adaptable to changing local climates (e.g., precipitation changes, colder or warmer weather, etc.) and able to contribute to broader efforts to mitigate the causes of climate change (e.g., greenhouse gas, or GHG, emissions). If we do not integrate climate change into these infrastructure developments now, there is a significant risk that we will be locked into fiscally and environmentally unsustainable infrastructure systems. As a recent Organisation for Economic Co-operation and Development (OECD) report put it,

> Choices made today about types, features and location of new and renovated infrastructure will lock-in "commitments" to future levels of climate change and to vulnerability or climate-resilience. Infrastructure vulnerability and risk to inevitable climate change is driven by long operational lifetimes of these investments, making them sensitive not only to the climate existing at the time of their construction, but also to climate variations over the upcoming decades. (Corfee-Morlot et al. 2012, 7)

It is critical, therefore, to integrate climate change and sustainable practices into the planning, design, construction, maintenance, and renewal of infrastructure systems before these long-term investments are made. This chapter is a contribution to this discussion. It draws on research undertaken over the last year and a half looking at whether and how climate change has been integrated into transportation infrastructure planning, design, and development in the province of Ontario.[1]

The chapter's primary purpose is to outline the implications of these issues for the engineering profession in order to enable engineers, and society more widely, to engage proactively with future infrastructure needs and choices. As such, it contributes to a growing debate about the role of engineers when it comes to climate change; this debate is evident in a range of initiatives, such as the British coalition Engineering the Future (Mair and Warry 2010); the establishment of the Public Infrastructure Engineering Vulnerability Committee (PIEVC) in 2005 by Engineers Canada (Lapp 2011a); the more recent survey of engineering practice by Engineers Canada (CSA Group 2012); and the new course, Adapting Your Infrastructure to Climate Change, organized by the Canadian Standards Association and Federation of Canadian Municipalities (FCM). The chapter is structured as follows: first, it will outline what we mean by the term *sustainable infrastructure*; second, it will provide some background on the research project on which this chapter is based; third, it will outline current examples of how climate change is being integrated into infrastructure in Canada and Ontario; and fourth, it will report on some of the findings about the implications of climate change for engineers.

CLIMATE CHANGE AND SUSTAINABLE INFRASTRUCTURE

In our research, we specifically used the term *infrastructure system* because infrastructure like roads, bridges, public transit, water, etc. represents more than isolated, physical artefacts. We adopted this perspective to align our research with a growing body of academic literature that emphasizes that adapting infrastructure to climate change entails more than just technical questions of, for example, load, usage, and deterioration; it also encompasses questions of social relevance, political choice, and economic cost (Geels 2004; Adger, Arnell, and Tompkins 2005; Shove and Walker 2007; Monstadt 2009; Brown, Furneaux, and Gudmundsson 2012; Frantzeskaki and Loorbach 2010; Markard 2011; Lawhon and Murphy 2012; Truffer and Coenen 2012; Birch and Wudrich 2013; McCormick et al. 2013). It is therefore helpful to think of infrastructure as an interdependent, social, and technical system or a socio-technical system (Hughes 1983; Graham and Marvin 2001; Markard 2011).

Transportation infrastructure provides a useful illustration of what we mean by a socio-technical system. The building of roads as a means of transport, for example, goes beyond the technical design of the road and the materials used during construction. To understand roads as a system requires that we examine the interdependent relationship between road building and (1) the manufacture of automobiles; (2) the extraction, processing, and distribution of oil; and (3) the social and/or political choices made available by these forms of personal transport (e.g., suburban living) (Lawhon and Murphy 2012). All these things, and more, form part of an infrastructure system that enables people to move from A to B and that is tied, implicitly, to questions of why people want to move from A to B (e.g., commuting, commerce, etc.). What this means is that any attempt to integrate climate change into infrastructure development has to address several interrelated yet, at the same time, seemingly distinctive issues (Shove and Walker 2007). Consequently, we cannot simply identify a single technical or engineering solution to climate change and the consequences of climate change for a given infrastructure system.

First, there are technical issues associated with the physical characteristics of infrastructure. These issues include the expected lifespan of an infrastructure system; its maintenance and repair; and its predicted (and actual) load, use, and capacity. Climate change may entail significant changes in load and usage, or increasing physical degradation as the result of weather changes (Goodings and Karney 2009). Second, there are social and political issues associated with the locational and/or distributional characteristics of infrastructure. They include decisions and choices about things like lifestyles and their relationship to density (e.g., suburbs, condos), population flows and their relationship to transit and transport (e.g., commuter congestion, public transit spending), and future town and regional planning (e.g., economic growth) (Truffer and Coenen 2012). Third, there are financial and economic issues associated with cost and capital spending to consider. These range from government procurement systems (e.g., direct public investment versus public-private partnerships) and the geographical scale of fiscal responsibility (e.g., federal versus municipal) all the way to individual household decisions about housing purchases. Consequently, it is possible to identify a range of significant economic and financial barriers to the integration of climate change into infrastructure (Monstadt 2009).

What we want to emphasize in this chapter is that sustainable infrastructure is our way of characterizing infrastructure that integrates climate change into its planning, design, development, and so forth; in this sense, we want to go beyond simple bricks and mortar (Birch and Wudrich 2013). The integration of climate change into infrastructure systems involves more than new forms of energy efficiency, smart technologies, and similar. Consequently, we argue that sustainable infrastructure has three key characteristics: first, infrastructure is an interdependent system in which social, political, and economic context matter as much as technical detail; second, any infrastructure change involves social and technological decisions and choices, not one or the other; and third, sustainable infrastructure entails forms of climate change mitigation (e.g., reducing GHG emissions) and, just as important, climate change adaptation by building in resilience to environmental changes (Adger, Arnell, and Tompkins 2005).

RESEARCH BACKGROUND

The research used in this chapter started in 2012 and is a collaboration between the authors and the Ontario Centre for Engineering and Public Policy (OCEPP), which is part of Professional Engineers Ontario (PEO). The aim of the research was to address four research questions.

1. What is the impact of climate change on infrastructure planning and development?
2. How is climate change integrated into infrastructure planning and development?
3. Are there barriers to integrating climate change into infrastructure planning and development?
4. What are the implications of this integration for the engineering profession?

We used a two-stage methodological approach. First, we sought to identify and map out a range of policy strategies and their implementation in Canada, Ontario, and Toronto. Second, we carried out in-depth interviews with engineers and associated professionals (e.g., architects, planners, etc.) involved in three transportation projects in Ontario. We chose transportation infrastructure because it has a direct impact on climate-change-mitigation strategies (e.g., GHG emissions reductions) and is directly impacted by changing climate (e.g., flood risks). Since infrastructure development has a life cycle, we chose three projects at different stages of the life cycle in order to capture the different stages of project development. We chose one project in the planning and design stage, a second project in the construction stage, and a third project in the renewal or repair stage. We also interviewed people from relevant organizations and institutions who are not directly involved in the three projects (e.g., local and municipal governments, professional associations, etc.), but who contribute to the broader policy environment.

In total, we interviewed 29 people. We asked everyone the same set of questions, which related to the four research questions listed above. We were particularly interested in exploring the following: whether climate change mitigation and adaptation were raised as concerns when it came to the specific projects and when it came to transport infrastructure generally; who raised mitigation and adaptation as concerns; if they were raised as concerns, how were they integrated into the planning, design, development, construction, and renewal of the specific project; what barriers had people encountered in integrating them; and what implications all of these issues had for engineers. Presented here are our findings about the implications of climate change and sustainable infrastructure for the engineering profession.

CANADIAN AND ONTARIO CONTEXT

Before we address the implications of climate change, however, we want to map out examples of current policy strategies, and their implementation, that promote sustainable infrastructure in Ontario and Canada. In limiting our focus to Canada and Ontario, we are not implying that these jurisdictions are ahead of the curve; we would, in fact, argue that both Canada and Ontario are actually lagging behind other places around the world (Mees and Driessen 2011). The integration of sustainability into infrastructure developments is evident in many other parts of the world, especially in the voluntary standards that have become increasingly popular over the last few years. Examples of these include a range of sustainable building certification and rating systems, some of which we list below (Kimmett 2012).

- BREEAM (BRE Environmental Assessment Method): established in 1990 in the United Kingdom, now very popular across Europe.

- LEED (Leadership in Energy and Environmental Design): established in 2006 in the United States; now an international standard.
- ecoENERGY: established by Natural Resources Canada in 2006; discontinued in 2012.
- HES (Home Energy Score): established in 2010 in the United States.

Most of these certification and ratings systems, however, are primarily concerned with resource and energy efficiency (i.e., mitigation) rather than sustainable infrastructure as we define it above. In this chapter, then, we are more interested in how climate change mitigation and adaptation are integrated into infrastructure developments. We have identified a number of examples of this integration in Canada and Ontario by mapping some of the policy debates in this area. To emphasize, sustainable infrastructure is not a new concept; rather, it is a term we are using to identify and define the policies, practices, standards, codes, etc. that are integrating climate change into infrastructure development. We sought to map out these examples of sustainable infrastructure planning, design, and development in order to understand better their implications for engineers, a topic that we return to in the next section. Below we outline some examples of policies in Canada; this is only an indication of the range of policies and is by no means a complete list.

Policy Strategies

Policy strategies in Canada are interesting for a number of reasons. First, national-scale strategies are limited to non-governmental organizations or agencies after the mid-2000s. While the Canadian federal government produced two important early strategies (Infrastructure Canada 2006; Natural Resources Canada 2007), they were not followed up with implementation or certification programs. Both strategies focused on the need to take adaptation seriously in policy-making. Alongside the federal government, other national agencies and organizations have produced their own strategies: (1) Engineers Canada (2008) has been very active in assessing the vulnerability of infrastructure to climate change; (2) the now defunct National Round Table on the Environment and the Economy produced a report on the need to mainstream adaptation into policy-making (NRTEE 2009); and (3) the FCM has produced at least two recent reports that address the implications of climate change for infrastructure (FCM 2011, 2012).

More recently, Canadian provincial and municipal governments have been more active in creating strategies than the federal government. For example, the Ontario government established an Expert Panel on Climate Change Adaptation in 2007, which produced a report in 2009. Since then, the Ontario government has produced a climate change action plan (2011–14) (Ontario 2011). Similarly, the city of Toronto produced a Climate Change, Clean Air and Sustainable Energy Action Plan in 2007, which led to its 2008 Climate Change Adaptation Strategy. The aim here is to improve information, especially weather data, and develop risk assessment tools to deal with change.

Policy Implementation

The above examples should show that there is mixed evidence of policy strategies at the federal, provincial, and municipal scales in Canada; unsurprisingly, this is similar when it comes to the implementation of climate change policies. At the national scale, the federal government created specific funding mechanisms to address climate change in the early 2000s; examples include the Canadian Strategic Infrastructure Fund (established in 2003) and Gas Tax Fund (established in 2005). Outside government, there has been more activity. For example, in 2005 Engineers Canada (2008) set up the PIEVC, which developed an Infrastructure Climate Risk Protocol (2008) to assess the impacts of climate change on infrastructure.

Generally, however, there has been more activity at the provincial scale. A few examples from Ontario include the Provincial Policy Statement (2014), which set out the requirements to consider the impacts of climate change and promote green infrastructure (sections 1.1.1, 1.6.1, and 1.6.2); the Ontario Building Code (2012 amendment), which increased its focus on energy efficiency; and the ten-year capital infrastructure plan, published in 2011, which was meant to take climate change into account in infrastructure development and ongoing asset management. Like Ontario, the city of Toronto has also implemented a number of policies to address climate change, including the Toronto Wet Weather Flow Master Plan (established in 2003); Toronto Green Standard (Toronto 2010), which is a series of performance measures for sustainable development; and the recent adoption of a Climate Change Adaptation Toolkit in 2013.

What this range of policy strategies and their implementation indicates is that climate change has been identified as an important issue at all levels of Canadian government, as have attempts to integrate climate change into various aspects of infrastructure development. It shows that climate change adaptation is being integrated into such developments alongside climate change mitigation. However, what it also shows is that these strategies and their implementation are relatively recent phenomena and are not necessarily coordinated or integrated across different scales of government.

IMPLICATIONS OF CLIMATE CHANGE AND SUSTAINABLE INFRASTRUCTURE FOR ENGINEERS

It would seem evident that the future and long-term planning, design, development, and maintenance of sustainable infrastructure depend on the engineering profession. We think there is a twofold challenge here for the profession. First, engineers are necessary for the technical success of sustainable infrastructure – they will play a crucial role in integrating climate change into infrastructure; and second, our society's need for sustainable infrastructure entails some significant challenges for engineers. Our perspective in this chapter is that the promotion and development of sustainable infrastructure will have benefits for Canada and Ontario as well as the rest of the world. As such, we think that it is necessary for engineers to face these challenges head on – and to do so now, while societies still have the opportunity to avoid being locked into unsustainable infrastructure. From the interviews we carried out with engineers and associated professionals, we identified five areas where climate change and sustainable infrastructure have significant implications for engineers; in the sections that follow, we go through each area in turn and raise several questions that we think are worthwhile to address.

Education and Continuing Professional Development

According to a recent survey of engineers carried out for Engineers Canada, a significant majority (74 percent) of infrastructure engineers reported that "they need more information to address the impacts of a changing climate in their practice" (CSA Group 2012, 12). While this may not be a surprising finding, it does have important implications for the education and continuing professional development of engineers.

First, if climate change is going to be integrated into infrastructure development, then engineering students need to be introduced to it as part of their undergraduate university education. Climate change needs to be built into the education and career development of engineers rather than becoming an afterthought or add-on to infrastructure development. Moreover, it is important to stress that addressing climate change involves more than questions of efficiency; adaptation and resilience are two concepts that need to be clearly incorporated into university programs. At present, it is not clear how many Canadian university engineering programs include courses on climate change and/or sustainability, although some do. For example, some universities do actively incorporate sustainability into their programs, although whether this includes issues of adaptation and resilience is unclear. One example is McMaster University, which has attempted to integrate sustainability across its engineering departments (McMaster University 2009).

Second, if universities are going to introduce students to climate change, then courses need to address more than just the technical aspects of infrastructure development (e.g., resource and energy efficiency). They also need to include the social dimensions of infrastructure systems (e.g., societal pressures, political decisions, and economic impacts) and learning outcomes that show that there is a range of possible solutions when designing and building infrastructure systems. There are already suggestions about core education aspects, including Engineers Canada's PIEVC, although this is more for professional associations than universities (Lapp 2011b).

Third, professional associations, both national and provincial, play an important role in promoting sustainable infrastructure through their continuing professional development (CPD) requirements. In fact, these CPD requirements are currently the only way to extend awareness of, and best practices for, climate change adaptation to most engineers (i.e., working engineers). It would therefore be helpful to coordinate such CPD requirements across Canada's professional associations in order to make the training consistent and coherent.

Finally, there is a perception that engineers are sometimes unwilling to accept that climate change is an important issue, one that they must address specifically themselves. This has more to do with a general concern about managing uncertainty (see "Professional Liability" below) and changing professional practices (see "Workplace Practices" below) than climate change skepticism; both of these concerns have costs associated with them, which have to be borne by someone, and they entail challenges to existing practices (e.g., professional liability, equipment familiarity, cost, etc.). Again, these implications are not simply a matter of education; they will involve a concerted effort to change the culture of engineering as much as its practice.

Workplace Practices

While there is a wide range of engineering practices, what we are focusing on in this chapter are workplace practices in the infrastructure sector that have implications for climate change. None of this is necessarily a new issue. The Canadian Society for Civil Engineering, for example, developed guidelines on sustainable engineering practices back in 1993, which it updated in 2006 (Mulligan 2011). What we want to stress here is that climate change and workplace practices are closely connected. This was evident in the interviews we carried out, especially in discussions about the need for, and benefits of, training and professional development in order to increase awareness and the capacity to respond to climate change. Thus, on-the-job practices are likely to be significantly impacted by the demands of sustainable infrastructure. We discuss four of these below.

First, one frequent comment we encountered when interviewing engineers was that good engineering was, by its very nature, sustainable because it involved the efficient use of resources (and sometimes energy). While this may be accurate, it is mostly relevant only when it comes to climate change mitigation; it is not as relevant when trying to integrate adaptation into infrastructure (e.g., weather resilience). This is a potential valuation of efficiency (resources and energy) over adaptability or resiliency. As these are all desirable outcomes, complexity theory may provide insights into a different prioritization of designs or outcomes (Kurtz and Snowden 2003). It is necessary to go beyond existing assumptions about engineering practice if we want to encourage sustainable infrastructure. Engineering practice needs to be green and resilient.

Second, integrating climate change adaptation into infrastructure development requires new information and risk assessment tools that are not, unfortunately, going to provide certainty (see "Professional Liability" below). As Mair and Warry (2010, 11), who also contributed to the *Infrastructure, Engineering and Climate Change Adaptation* report for the Engineering the Future alliance, put it, "The effects of climate change are not easy to predict. Engineers need to embrace probabilistic methods and flexible solutions, and must be able to deal with complex risk scenarios which involves a range of factors." New tools are needed to alleviate existing information deficits when it comes to things like weather and climate patterns; however, incorporating these tools into existing practice might require new approaches in the workplace.

Third, it is obvious that engineering practices are strongly driven by client demands and especially financial considerations, both of which are perhaps self-evident. However, engineers can play a much greater role in promoting sustainable infrastructure by highlighting the cost savings of sustainability measures (e.g., energy savings) over the life cycle of an infrastructure project (Holtforster and Nielsen 2011). Actively pushing for the inclusion of such measures would be one way to challenge systemic barriers to climate change integration (e.g., public-procurement bidding processes). One thing is clear: engineers and asset or financial managers need to engage in greater dialogue with one another to promote sustainable infrastructure.

Finally, one important issue that is rarely addressed is the need to update construction equipment in response to climate change; however, new equipment will have an impact on workplace practices because it is likely to entail significant changes and expense. First, it is costly to replace equipment; second, it is costly to retrain people to use it. The likely increase in cost is a major barrier since it will deter developers from paying for changes to infrastructure.

Professional Liability

During our project, we found that interviewees laid particular stress on the public responsibility of engineers; it was seen not as an afterthought but, rather, central to their work. As the PEO's *Code of Ethics* states, "A practitioner shall … regard the practitioner's duty to public welfare as paramount" (PEO 2012, ss. 2). Obviously, how we define *public welfare* is critical for understanding the implications of climate change and sustainable infrastructure for engineers now and in the future. Another important consideration is that, in light of changing climate and weather, professional (and personal) liability is a more immediate concern for engineers than future public welfare.

First, engineers stressed their (ethical and professional) responsibility to the public in their role, especially when it comes to infrastructure development. While public welfare or the public interest can be defined in a number of ways (e.g., health, wealth, happiness), it seems that engineers understand it in a particular way: primarily to promote and ensure safe designs for infrastructure (see "Codes and Standards" below). This laudable goal is tied, inextricably, to the information that engineers have available to them about future demands on the infrastructure they have designed. Consequently, since information about future weather and climate patterns is (inherently) uncertain, engineers face significant challenges in developing infrastructure that will be adaptable and resilient as well as in the public interest.

Second, what seemed to be the most pressing concern for engineers was safety; prosaically, infrastructure should not collapse. One key connection between safety and liability is the issue of negligence in design; this could include the use of outdated engineering practices or information. It is evident that climate change presents a significant challenge here since it is increasingly difficult to know how (local) weather and climate will impact infrastructure (or even what the weather and climate will be). Being willing and able to incorporate new risk assessment or modelling tools seems crucial in this regard, even when information may be uncertain (see "Workplace Practices" above).

Third, those we interviewed seemed to have a general perception that engineers are not adept at managing the increasing uncertainty of climate benchmarks, such as rainfall and temperature, and that practices and education need to change as a result. More is needed, however, as being able to manage uncertainty involves engineers making an attitudinal, cultural, and legal shift. How this might be supported is complicated, but it probably requires addressing several areas of engineering practice in tandem (education, workplace practices, and codes). On a practical level, it would seem that this issue will create the greatest difficulties for engineers since they are more used to working from immutable physical laws and stable historical data.

Finally, professional liability is an interesting issue with respect to climate change and engineering because of the uncertainty about the future demands or pressures on infrastructure (e.g., changing weather patterns, changing lifestyles). While uncertainty may preclude some design decisions, it does not stop engineers from thinking about the adaptability and resilience of the infrastructure they design. It is possible that ignoring uncertainty will not be a good enough defence against negligence in the future; this means that over-designing infrastructure, based on existing standards, may be a responsible course of action at present. What is evident, however, is that there probably needs to be a legal motivation for including climate change in engineering practice.

Codes and Standards

Engineers are constrained and driven by codes and standards; at least, that is the impression we gained from our interviews. Examples of how codes and standards have promoted or can promote resource and energy efficiency are not difficult to find, especially when it comes to buildings and GHG emissions reduction. As Sugar and Kennedy (2013) point out, working to specific standards and building codes (e.g., Ontario Building Code, Toronto Green Standard) can lead to significant reductions in GHG emissions, as can promoting certain kinds of transport infrastructure (e.g., public transit, cycling). What is less clear, however, is how codes and standards can promote sustainable infrastructure as we define it, especially the adaptability and resilience of infrastructure to changing climates. This is, inherently, far more difficult to address in codes and standards because it relates to future changes independent of buildings and/or infrastructure.

First, engineers are driven by codes and standards in their work; these are, in fact, foundational to how engineers work. It is, therefore, important to note that the engineering profession will never be able, by and of itself, to promote sustainable infrastructure or other ways of addressing climate change without changes to the applicable codes and standards. There is a significant mismatch here between the work that engineers can do and what we might want engineers to do if we wish to actively promote the integration of climate change into infrastructure development. It is a classic chicken-and-egg situation: to achieve a change, either engineers can change their practices to influence codes or codes can be changed to influence engineering practice.

Second, it is not clear when and where a problem like climate change should (or whether it can) be integrated into the design process; it could be at the planning stage, or design, or construction, etc. More reasonably, it is probably necessary to integrate it into every stage, but this would entail a progressive change in codes and standards throughout the infrastructure life cycle – from zoning to building codes to energy efficiency requirements for equipment and so on.

Third, one key problem with promoting adaptability and resilience is that codes are often based on established, historical data – largely out of necessity and comparability to engineering fields that use verifiable material characteristics. In this sense, codes derived from weather or climatic data react to events rather than driving the promotion of particular goals (e.g., sustainability, behavioural change, etc.). This means that attempts to integrate adaptation into infrastructure development might be possible only if engineering codes and standards can be informed by probabilistic models and changing the balance between efficiency and sustainability in engineering designs and decisions.

Finally, engineers can and do assist in the development of codes and standards. It was evident that interviewees thought it important for engineers to be more actively involved in developing codes and standards to include a heavier weighting toward adaptation goals. The time frame for developing new codes and standards can be long, however; it can take several years for changes to be introduced and then have an influence. An example of this is the Toronto Green Standard; it was made mandatory in 2010, but its history stretches back to 2004.

Employment and Jobs

There is a worldwide infrastructure challenge resulting from the need to build new or renew aging infrastructure. According to an OECD report, this will cost trillions of dollars by 2030 (Corfee-Morlot et al. 2012). Similar challenges exist in Canada, where significant under-investment in infrastructure has occurred since 1980; as a result, aging infrastructure is reaching its end of life, compounding the challenge and resulting in a municipal infrastructure deficit of $123 billion and the need for $115 billion of new infrastructure investment (FCM 2007). What these infrastructure needs highlight is that there is plenty of work for engineers to undertake over the next decades, and it will therefore continue to be a thriving profession.

First, while there is potentially a lot of work for engineers in the coming decades, it is likely to appear only if the necessary public investment is made. This is not certain in the current political and economic environment (CCA 2011). The financial cost, public willingness to pay, and political inertia all represent major barriers to sustainable infrastructure; lack of public investment, whether for sustainable infrastructure or not, is likely to impact the number of jobs available for engineers. While market-based funding mechanisms (e.g., public-private partnerships) have become increasingly popular, they represent only a small proportion of total infrastructure investment; they are therefore unlikely to make up any shortfall.

Second, most interviewees believed that sustainable infrastructure will necessitate more work at almost all stages of the infrastructure life cycle, leading to more jobs for engineers. While this was the majority opinion, it is worth noting the minority view: that integrating climate change into infrastructure implies changing education and work practices, but not more work per se. It could involve a wholesale shift, for example, in the time horizons that engineers work to rather than the extent of work itself.

However, that being said, it is more likely that promoting sustainable infrastructure will lead to new work opportunities and therefore more engineering employment.

Third, no country has taken a definitive lead in building sustainable infrastructure. While some countries are more active than others in integrating climate change into infrastructure development, there is less evidence that this has involved integrating adaptation alongside mitigation. Engineers in Canada have the opportunity to take the lead on this sort of development; this could mean rising employment and more jobs for engineers trained in sustainable infrastructure development. However, any such opportunity comes with certain challenges, which would include a significant change across the board for the engineering profession (see "Education and Continuing Professional Development," "Workplace Practices," "Professional Liability," and "Codes and Standards" above).

Finally, sustainable infrastructure may offer employment opportunities for engineers, but it could also threaten other jobs. One example raised by an interviewee was the threat posed by new work practices and equipment to levels of employment in the construction sector. It is also likely, considering the systemic interdependencies in dealing with global problems like climate change, that some engineering sectors will be threatened by a major turn toward climate change mitigation and adaptation in infrastructure (e.g., petroleum engineering, automobile engineering, etc.). If climate change is not dealt with systematically, however, it is likely going to be difficult to address in a meaningful manner. What this means is that educational programs and courses may have to change in order to promote certain engineering sectors at the expense of others (see "Education and Continuing Professional Development" above).

CONCLUSIONS

In this chapter, we have attempted to highlight some of the issues raised by engineers and other relevant stakeholders in infrastructure development. While we have focused on Ontario and Canada, we think that much of what we have written is relevant for other parts of the world. This chapter is really about identifying existing examples of what we have termed *sustainable infrastructure* and then considering the implications of this infrastructure for the engineering profession. What we mean by sustainable infrastructure is infrastructure that integrates climate change mitigation and adaptation into the planning, design, construction, and maintenance phases of its life cycle. Moreover, it is meant to stress the social and technical aspects of infrastructure by highlighting how engineering practices are bound up with other considerations.

While integrating climate change into infrastructure development has a range of implications for engineers, as we have outlined above, it is an urgent issue in order to avoid the continued use (i.e., lock-in) of unsustainable technologies, processes, practices, and standards. Engineers must actively engage in the transformation of their educational and work practices. However, they face a number of specific professional challenges that are difficult to resolve, including how engineering practice is driven by historical data; this places severe limitations on the development of sustainable infrastructure. Nevertheless, the people we interviewed generally agreed that there is a need for a societal shift toward addressing climate change.

Generally, climate change is a complex and global problem, and it does not look like coordinated mitigation efforts are going to solve it any time soon. As a result, climate adaptability and resilience are increasingly important strategies; this is especially the case when it comes to infrastructure since most countries, including Canada, need major overhauls to their existing infrastructure and have an ongoing need to build new infrastructure. If we do not try to promote adaptability and resilience now, it is very possible that our infrastructure will be at risk in 20 or 30 years' time, before the end of its useful life. Moreover, it is probable that we will become locked into the continuing use of unsustainable practices and policies. Adapting the engineering profession to integrate climate change is an urgent issue, but one that we think engineers are very well placed to address.

NOTE

1. The project was funded as part of the Work in a Warming World Community-University Research Alliance (2010–15) of the Social Sciences and Humanities Research Council of Canada.

REFERENCES

Adger, W.N., N.W. Arnell, and E.L. Tompkins. 2005. "Successful Adaptation to Climate Change across Scales." *Global Environmental Change* 15 (2):77–86.

Birch, K., and D. Wudrich. 2013. "Climate Change, Sustainable Infrastructure and the Challenge Facing Engineers." *Engineering Dimensions* (September/October):46–48.

Brown, K., C. Furneaux, and A. Gudmundsson. 2012. "Infrastructure Transitions towards Sustainability: A Complex Adaptive Systems Perspective." *International Journal of Sustainable Development* 15 (1/2):54.

CCA (Canadian Construction Association). 2011. "Infrastructure in a Post-stimulus World." Paper presented at the Public Policy Conference, Toronto, 6 May.

Corfee-Morlot, J., V. Marchal, C. Kauffmann, C. Kennedy, F. Stewart, C. Kaminker, and G. Ang. 2012. "Towards a Green Investment Policy Framework: The Case of Low-Carbon, Climate-Resilient Infrastructure." OECD Environment Working Paper 48. Paris: OECD Publishing.

CSA Group. 2012. "National Survey of Canada's Infrastructure Engineers about Climate Change." Ottawa: Engineers Canada.

Deveau, D. 2012. "Cities Breaking Down Departmental Barriers to Improve Future Infrastructure." *Financial Post*, 10 April. http://business.financialpost.com/2012/04/10/a-new-approach-to-infrastructure/?_lsa=8a6b-e94c.

Engineers Canada. 2008. "Adapting to Climate Change: Canada's First National Engineering Vulnerability Assessment of Public Infrastructure." Ottawa: Engineers Canada.

FCM (Federation of Canadian Municipalities). 2007. "Danger Ahead: The Coming Collapse of Canada's Municipal Infrastructure." Ottawa: FCM.

—. 2011. "Building Canada's Green Economy." Ottawa: FCM.

—. 2012. "The Road to Jobs and Growth." Ottawa: FCM.

Frantzeskaki, N., and D. Loorbach. 2010. "Towards Governing Infrasystem Transitions: Reinforcing Lock-In or Facilitating Change?" *Technological Forecasting and Social Change* 77 (8):1292–1301.

Geels, F. 2004. "From Sectoral Systems of Innovation to Socio-technical Systems: Insights about Dynamics and Change from Sociology and Institutional Theory." *Research Policy* 33 (6/7):897–920.

Goodings, D., and B. Karney. 2009. "Climate Change Policy: Beyond the Mechanics." Presentation at OCEPP Policy Engagement Series, Toronto, 8 May.

Graham, S., and S. Marvin. 2001. *Splintering Urbanism*. London: Routledge.

Holtforster, F., and R. Nielsen. 2011. "When It Comes to New Buildings, Sustainability Pays." *Engineering Dimensions* (March/April):52–54.

Hughes, T. 1983. *Networks of Power: Electrification in Western Society, 1880–1930*. Baltimore: Johns Hopkins University Press.

Infrastructure Canada. 2006. "Adapting Infrastructure to Climate Change in Canada's Cities and Communities." Ottawa: Infrastructure Canada.

Justian, E. 2013. "Failing Infrastructure Could Cost U.S. Businesses $1.2 Trillion by 2020." *Triple Pundit*, 3 June. http://www.triplepundit.com/2013/06/failing-infrastructure-could-cost-businesses-12-trillion-2020/.

Kimmett, C. 2012. "Get More Mileage from Your Home." *Alternatives Journal* 38 (2):32–34.

Kurtz, C., and D. Snowden. 2003. "The New Dynamics of Strategy: Sense-Making in a Complex and Complicated World." *IBM Systems Journal* 42 (3):462–83.

Lapp, D. 2011a. "Adapting Civil Infrastructure to a Changing Climate." Presentation at the OCEPP Policy Engagement Series, Toronto, 10 February.

—. 2011b. "Climate Change and Engineering: Education and Continuing Professional Development Initiatives." Presentation at the National Academy of Engineering Workshop as part of the "Networking Educational Priorities for Climate, Engineered Systems, and Society" project, Washington, DC, 19 October. http://www.pievc.ca/e/nae_climate_change_and_education.pdf.

Lawhon, M., and J. Murphy. 2012. "Socio-technical Regimes and Sustainability Transitions: Insights from Political Ecology." *Progress in Human Geography* 36 (3):354–78.

Mair, R., and P. Warry. 2010. "Opinion: Infrastructure, Engineering and Climate Change." *Ingenia* 45 (December):10–11.

Markard, J. 2011. "Transformation of Infrastructures: Sector Characteristics and Implications for Fundamental Change." *Journal of Infrastructure Systems* 17 (3):107–17.

McCormick, K., S. Anderberg, L. Coenen, and L. Neij. 2013. "Advancing Sustainable Urban Transformation." *Journal of Cleaner Production* 50:1–11.

McLaughlin, M. 2013. "Change in the Weather." *PE Magazine* (March). http://www.nspe.org/PEmagazine/13/pe_0313_Change.html.

McMaster University. 2009. *Engineering a Sustainable Society: Strategic Plan 2009–2014*. Hamilton, ON: Faculty of Engineering, McMaster University. http://www.eng.mcmaster.ca/strategicplan/StrategicPlan.pdf.

Mees, H.-L., and P. Driessen. 2011. "Adaptation to Climate Change in Urban Areas: Climate-Greening London, Rotterdam, and Toronto." *Climate Law* 2 (2):251–80.

Monstadt, J. 2009. "Conceptualizing the Political Ecology of Urban Infrastructures: Insights from Technology and Urban Studies." *Environment and Planning A* 41 (8):1924–42.

Mulligan, C. 2011. "CSCE Guidelines for Sustainable Development." *Engineering Dimensions* 32 (2):51–61.

Natural Resources Canada. 2007. "From Impacts to Adaptation: Canada in a Changing Climate." Ottawa: Natural Resources Canada.

NRTEE (National Round Table on the Environment and the Economy). 2009. "True North: Adapting Infrastructure to Climate Change in Northern Canada." Ottawa: NRTEE.

Ontario. 2011. *Climate Ready: Ontario's Adaptation Strategy and Action Plan 2011–2014*. Toronto: Government of Ontario.

PEO (Professional Engineers of Ontario). 2012. *Code of Ethics*. S. 77 of the O. Reg. 941. http://peo.on.ca/index.php?ci_id=1815&la_id=1.

Shove, E., and G. Walker. 2007. "CAUTION! Transitions Ahead: Politics, Practice, and Sustainable Transition Management." *Environment and Planning A* 39 (4):763–70.

Stern, N. 2006. *Stern Review: The Economics of Climate Change*. London: HM Treasury. http://webarchive.nationalarchives.gov.uk/20100407172811/http://www.hm-treasury.gov.uk/stern_review_report.htm.

Sugar, L., and C. Kennedy. 2013. "A Low Carbon Infrastructure Plan for Toronto, Canada." *Canadian Journal of Civil Engineering* 40 (1):86–96.

Toronto. 2010. "Toronto Green Standard: Making a Sustainable City Happen." Toronto: City of Toronto. https://www1.toronto.ca/city_of_toronto/city_planning/developing_toronto/files/pdf/lr_res_tech.pdf.

Truffer, B., and L. Coenen. 2012. "Environmental Innovation and Sustainability Transitions in Regional Studies." *Regional Studies* 46 (1):1–21.

Wells, J. 2012. "Climate Change: How Toronto Is Adapting to Our Scary New Reality." *Thestar.com*, 19 August. http://www.thestar.com/news/insight/2012/08/19/climate_change_how_toronto_is_adapting_to_our_scary_new_reality.html.

CHAPTER 7

CONSTRUCTION AND CLIMATE CHANGE: OVERCOMING ROADBLOCKS TO ACHIEVING GREEN WORKFORCE COMPETENCIES

JOHN CALVERT

INTRODUCTION

The critical role of labour in addressing climate change has been neglected in recent research. While the expanding literature on climate issues has identified a wide range of measures to mitigate and, subsequently, to adapt to the projected rise in global temperatures, it has generally taken for granted that the workforce will be able to implement such measures effectively. But this is not a given. The Canadian construction sector has the potential to make a key contribution to reducing greenhouse gas (GHG) emissions and energy use. However, it has major gaps in training and skills development that compromise its ability to do so. Canada's system of apprenticeship remains focused on traditional competencies while paying little attention to providing workers with the skill sets and knowledge necessary to implement low-carbon construction in the built environment. The increasing focus of the federal government – and a number of provinces – on the traditional training needs of resource extraction industries has further impeded this transition. Rather than adopting policies to ensure that construction workers have the skills to retrofit the existing building stock, and to build new, more energy-efficient structures, industry training investments are increasingly being targeted at the trades that oil, gas, and mining corporations need.

Work in a Warming World, edited by Carla Lipsig-Mummé and Stephen McBride. Kingston: School of Policy Studies, Queen's University.

Reversing these developments will not be easy. It will require governments to adopt a different policy agenda, one that challenges the market-driven approach currently shaping Canada's construction training system. It will also require a commitment to workforce development based on the understanding that low-carbon construction requires higher levels of training and a profound cultural change in the industry in which building workers – and the unions that represent them – view climate objectives as a fundamental part of their role in the industry.[1]

CONSTRUCTION AND GLOBAL WARMING

The construction industry is a major component of Canada's economy. It employed 1.3 million people and accounted for $113 billion of economic activity in 2012, or approximately 7.2 percent of the nation's gross domestic product (Statistics Canada 2013), and it employed 7.4 percent of the country's workforce at the beginning of 2013 (Statistics Canada 2014b). The industry and its workforce have grown considerably over the past decade despite the impact of the 2007–09 recession, and BuildForce Canada (formerly known as the Construction Sector Council, or CSC) predicts further significant expansion in the coming years.[2] However, the pattern of expansion has varied both by sector of construction and by geographic location. Much of this expansion has reflected Canada's increasing reliance on resource extraction industries, principally oil, gas, and mining, as well as related infrastructure developments. Geographically, the expansion has been most pronounced in Alberta, British Columbia, Saskatchewan, parts of the north, and Newfoundland and Labrador, with a projected total job gain of 44,000 building workers in non-residential construction in Canada by 2021. In contrast, residential construction investment and employment has remained – and is projected to remain – relatively flat, with a modest decline of 8,000 jobs by 2021 (CSC 2013).

Energy production in the tar sands and in the emerging gas-hydraulic-fracturing ("fracking") and related liquefied-natural-gas-processing industries, along with the infrastructure needed to support these activities, relies on a well-trained construction workforce. The tar sands can be described as one huge construction site. It employs large numbers of skilled building trades workers, some of whom have skills and training applicable to many different kinds of construction projects, while others have skills largely dedicated to working in the energy sector. Given their need for skilled trades workers, the large oil and gas corporations now play a major role in shaping federal- and provincial-government training priorities in the construction sector, including exercising considerable influence on many of the organizations that play a role in providing and overseeing training. They have also established strong ties with some of the key building trades unions whose members depend on employment in the resource sector.

What is perhaps most striking about these developments is the lack of concern by either the industry or governments about the long-term climate impacts of Canada's pattern of construction expansion and, particularly, the resulting qualifications profile of the construction workforce, with its growing dependency on skills used primarily in the resource extraction sector. Workforce training and skills development are being skewed toward servicing the demands of the resource industries, independent of any recognition of the need to reduce climate impacts in the one sector where gains are most promising: the built environment.[3] This is not a result of lack of information about the opportunities in the built environment. For the potential gains – both in climate change mitigation and employment creation – from aggressive investments in this sector of construction are well researched and well understood, around the world, as evidenced by the work carried out by the Intergovernmental Panel on Climate Change (IPCC), United Nations Environment Programme (UNEP), International Labour Organization (ILO), European Union, and numerous other organizations.

Domestically, BuildForce Canada, which has focused much of its attention on the needs of the energy sector, has also sponsored research on opportunities to implement low-carbon construction (CSC 2011). The construction industry's support for the provision of additional resource-focused training and skills development reflects the influence of the energy and mining corporations and their successful efforts to have the Canadian government base much of its economic strategy – and the training required to achieve this strategy – on expanding resource exports.

A fundamental transformation of Canada's construction industry focusing on green construction in the built environment is urgently needed. But this is contingent on larger changes in Canada's overall economic strategy, including reducing Canada's reliance on fossil fuel exports. In the present political climate, such changes appear unlikely, at least in the immediate future. However, in the longer term, and assuming that the scientific projections of incrementally rising global temperatures prove accurate, an economy that depends on continuous expansion of energy-intensive resource exports will be increasingly difficult to sustain. Whether sooner or later, governments – and the construction industry – will need to focus much more attention on what can be done to cut energy use and GHG emissions in the built environment.

It is the focus of this chapter to examine some of the challenges that the industry and, more specifically, its workforce currently face in making this transformation and to suggest some modest policy changes that will facilitate this process. Thus, the chapter does not address the big-picture issue of what will be needed to redirect the overall economic priorities of the federal and most provincial governments. Nor does it deal with the difficult question of how to supplant the political and economic priorities of the fossil fuel industry with a policy framework that takes account of the real threat that global warming poses to Canada.

Rather, it focuses, more narrowly, on questions associated with what needs to be done to facilitate the implementation of a training and skills development agenda that will make it easier to implement green construction in the built environment in the coming years. Clearly, implementing this narrower agenda will be contingent on the big-picture changes noted above. But even if or, perhaps more optimistically, when these big-picture changes take place, it will still be necessary to address the practical questions of how to ensure that the construction industry and, specifically, its workforce is up to the task of implementing low-carbon construction.

The reason this chapter focuses on the built environment is because of its significant potential to mitigate climate change. Buildings produce an estimated 30 to 35 percent of GHG emissions. They also account for almost 40 percent of energy use. A recent study by the ILO, which examined the potential for reducing GHG emissions and energy use in the world's major economies, concluded,

> Of all the elements that constitute society, buildings are the biggest consumers of energy and the largest emitters of greenhouse gases. Yet, the building sector also has the highest potential for improving energy efficiency and reducing emissions. Many investments in resource-efficient buildings are cost effective and the large stocks of older and inefficient buildings, notably in industrialized countries, mean that placing greater emphasis on renovation could yield substantial environmental benefits. (ILO 2012, 127)

Of course, the built environment consists of a wide variety of different structures and buildings, including individual residences, low- and high-rise apartments, commercial premises, office buildings, factories, industrial plants, hospitals, schools, public facilities, resource extraction projects, and numerous other elements. There is no simple, cookie-cutter template for reducing its climate impact. The challenge is complex. At a minimum, cutting energy use and GHG emissions in the built environment involves designing and introducing energy conservation systems in new buildings. More significantly, it also involves developing appropriate measures to facilitate the wholesale retrofitting of the existing building stock.

Over the past two decades, since concerns about climate change began to emerge, there has been growing interest in low-carbon construction among environmentally conscious developers, architects, engineers, general contractors, policy researchers, and climate activists. However, much of this interest has focused on new construction. In the long term, given the life expectancy of most new construction, ensuring that new building projects have a much-reduced climate footprint must be a major priority. However, retrofitting the existing building stock presents much larger opportunities for energy savings and GHG reductions in the near and mid-term time frame. In its 2007 examination of the building sector, the authors of an IPCC report on climate change mitigation noted, "Over the whole building stock the largest portion of carbon savings by 2030 is in retrofitting existing buildings and replacing energy-using equipment due to the slow turnover of the stock" (Levine and Ürge-Vorsatz 2007, 389).

This is because new construction represents only a very small proportion – about 1.5 percent – of the existing stock of buildings each year. Consequently, retrofitting existing buildings is critical to achieving effective climate change mitigation (UNEP 2009).[4] However, this is a major challenge because of the wide variations in the type, age, and quality of Canada's existing building stock. Effectively retrofitting the built environment is a particularly daunting task – one that will require ongoing investments in research, new technologies, and, from the narrower perspective of this chapter, extensive workforce training and skills development to give the industry – and its workers – the capacity to implement low-carbon construction on the job site. It will also require a major shift in public policy to facilitate a fundamental reorientation of the industry.

Of course, the focus of this chapter – implementing major changes in the skills profile of the workforce – can take place only in the context of a larger industry transformation; this involves every step in the supply chain – from the decisions made by those who commission and purchase new construction services and renovations; through the work of the architects, planners, and engineers who design (or renovate) the buildings; to the general contractors and subcontractors who oversee the projects; and, finally, to the skilled tradespeople and labourers who actually carry out the work on building sites.[5] At each stage of this process, the industry needs to implement the most effective and technologically advanced low-carbon construction innovations. And all those who work in the industry must be trained for and committed to the same objective. In the words of the IPCC report noted above, it will require "very significant efforts to enhance programmes and policies ... well beyond what is happening today" (Levine and Ürge-Vorsatz 2007, 390).

CANADA'S EXISTING TRAINING SYSTEM IS INCAPABLE OF TRAINING THE WORKFORCE IN LOW-CARBON CONSTRUCTION TECHNIQUES

A major impediment to achieving low-carbon construction is Canada's fragmented, under-resourced, and poorly regulated system of construction industry training. Put simply, the current training system and, consequently, the current workforce is not up to the job. It not only lacks the capacity to provide the green skills required for low-carbon construction; it also does not have the ability to provide adequate training for much of the work currently being done in the industry. An industry that fails to address its current training requirements is an industry that does not have a workforce capable of implementing green construction either, particularly because green construction requires higher levels of skills and training, a point to which we shall return later in this chapter.

Of course, criticisms of the construction industry's existing training system are not new. The weaknesses in Canada's training and apprenticeship system have been widely noted – not only by critics of the industry but also by many of its major players, including federal and provincial governments, developers, and numerous construction firms. It is not accidental that the industry has pushed the federal government to encourage recruitment of large numbers of skilled trades workers from other countries to fill the skills gap in Canada's construction workforce. At the beginning of 2014, for example, the BuildForce website provided no fewer than 14 publications to help employers recruit foreign trades workers – a reflection of the industry's unmet demand for skilled construction workers. (In comparison, the website provides one survey paper on greening residential construction.)

Virtually all construction in Canada, including that carried out in the built environment, is a private sector activity. The economic interests of the purchasers of construction services, the developers who commission these services, and the contractors who deliver these services shape and control virtually all aspects of what is built, who builds it, and how it is built. Even when government is the purchaser of construction services, private firms carry out the actual work.

This is not to deny that there is extensive public regulation of many aspects of construction projects. There are detailed building codes, fire safety regulations, electrical codes, plumbing standards, engineering requirements, and policies requiring the use of specific construction materials, to cite only a few of the numerous regulations in force. There is public oversight of planning and development by applying provincial and municipal zoning regulations, which influence – at least to some degree – what is built and where it is built. However, in practice, the construction industry, through the operation of the private market, exercises considerable influence over much of the public regulation that exists, ensuring that most regulation is industry friendly in that it accommodates the industry's focus on maximizing profits and minimizing regulations that would challenge this basic objective.

Consequently, widespread implementation of low-carbon objectives in new construction has been slow and erratic, depending largely on the market-driven preferences of developers and the willingness of purchasers to invest in energy conservation, helped along by modest regulatory changes brought in by various levels of government. Progress on renovating the existing building stock has been even slower, a reflection, in part, of the industry's market-based approach, which is strongly resistant to public planning and public regulation.[6]

It is true that some municipal planners, developers, architects, and engineers are attempting to introduce greener building practices. Municipalities and provinces have strengthened their building codes. There is now a variety of climate-friendly standards for the built environment, such as LEED (Leadership in Energy and Environmental Design), BOMA (Building Owners and Managers Association), Built Green, Green Globes, ISO 14001, and Passive House, to name some of the more widely used systems. But these initiatives are not mainstream. They are also voluntary. They rely on the decisions of building purchasers, owners, and developers, who, overwhelmingly, still base their choices on the perceived costs and benefits of such innovations. Moreover, being voluntary, there are also major questions about how effectively these innovative approaches are being implemented and monitored. And there are concerns that some of these innovations are little more than green sales and marketing strategies.

Regardless of the efforts of some in the industry to promote greener construction, there remains an enormous gap between what needs to be done and what the construction industry, as currently organized, can deliver. Competition, based almost exclusively on low bid, results in a race to the bottom as firms strive to obtain work by cutting corners on design, materials, technology, and, not inconsequentially, labour. Rather than upward pressure encouraging adoption of the highest environmental standards, unregulated price competition results in the opposite: pressure to adopt the minimum standards that will succeed in a competitive market (Barrett 1998; Bosch and Philips 2003; Prism Economics & Analysis 2010). While this focus may satisfy the economic interests of most of the industry as presently constituted, it does not provide the basis for low-carbon construction.[7]

Industry reluctance to implement greener construction is reinforced by a culture strongly committed to the free market and hence highly resistant to attempts by governments to introduce more directive public regulation, whether of planning, design requirements, engineering standards, tougher building codes, demolition protocols, new technologies, or mandated workforce training and skills qualifications.[8] The industry is normally willing to accept such changes only when they do not impose major costs, can be introduced with minimal disruption to its activities, and do not otherwise interfere with its profit-maximizing objectives (Bosch and Philips 2003).

Relying on free market principles means that the industry's labour market is also poorly regulated. Fluctuations in the business cycle, as well as seasonal weather changes, result in a boom-and-bust pattern of building activity and employment. These changes are much more dramatic than in most other sectors of Canada's economy, and they are exacerbated by the reluctance of governments to play a larger role in addressing fluctuations in the construction market through anti-cyclical policies. This makes building projects risky. The way the industry deals with risk is largely through subcontracting (Bosch and Philips 2003), a practice that enables developers and contractors to reduce their exposure to cyclical fluctuations by increasing, or decreasing, the volume of subcontracts as demand fluctuates. It also enables them to reduce the risk of equipment and facilities lying idle by shifting responsibility for capital investments to smaller operators. Because they do not employ their subcontractors' workers, developers and general contractors minimize their employment obligations, transferring the costs of downturns to smaller subcontractors, their employees, and the large number of owner-operators.[9]

Labour market flexibility is achieved largely at the expense of the workers, who face periodic and unforeseen unemployment, whether directly as employees or indirectly as self-employed owner-operators. Aside from creating income and work insecurity, precarious employment undermines long-term attachment to the industry. Instead of having a career, many workers – particularly young apprentices – are forced to abandon the industry during downturns to seek work in other sectors of the economy, from which they often do not return. High labour turnover means that the industry loses valuable skills and squanders a significant part of its – and its workers' – investment in training. It is a hugely wasteful and, arguably, socially irresponsible approach to labour force development.

The extensive use of subcontracting also results in a very fragmented industry structure. While there are a number of very large contractors, especially in resource extraction, infrastructure construction, and the industrial, commercial, and institutional (ICI) sector, the industry is overwhelmingly composed of very small firms or individual owner-operators. Fully 89 percent of employers have 20 or fewer workers, and 61 percent have fewer than five (O'Grady 2005; Hamilton-Smith 2010).[10] Even in larger projects, including the tar sands, extensive subcontracting is the norm.[11]

These various industry characteristics make it difficult to provide an effective training system. Small contractors and subcontractors normally do not have the qualified journeypersons or other resources to support apprentices. Larger firms with the capacity to do so are often reluctant because of the cost and the uncertainty of future work contracts. Moreover, there is nothing preventing competing contractors who have made no investment in training from poaching newly qualified workers. As we shall discuss further in this chapter, unions have addressed this problem by reaching multi-employer agreements, while Quebec has established a statutory framework that requires employers to support training. But this is an option that most of the industry in English Canada does not support. Outside Quebec, governments have not been prepared to require contractors to support training financially. Nor have they required them to take on apprentices. Government efforts to address skills shortages have generally used indirect approaches involving carrots such as subsidies to employers or tax breaks for workers, measures that have been largely ineffective (Hamilton-Smith 2010). Given this situation, it is perhaps not surprising that Canada's construction industry has not been very successful in developing a workforce capable of achieving the transition now required to cope with global warming.

BARRIERS TO EFFECTIVE GREEN TRAINING

While many factors have contributed to Canada's construction training crisis, it is possible to highlight three in particular that are major impediments to creating a workforce with appropriate green skills. The first is the impact of the underground economy. It drives down building standards, impedes the introduction of more energy-efficient building techniques, fails to provide workforce training, and encourages the use of unskilled labour. The underground economy is supported by the industry's extensive reliance on subcontracting, which facilitates small-scale, under-the-radar building activity. Government reluctance to regulate the underground economy allows its problematic practices to flourish.

The second factor is the inadequacy of the current training and apprenticeship system. This is an issue that the industry acknowledges but has failed to address effectively. Too many apprentices never find sufficient work to complete their training, and too many employers fail to contribute anything at all to supporting workforce training.

And the third factor is the widespread neglect of the potential role that workers and the unions that represent them could – and, arguably, should – play in the process of greening the built environment, a neglect based on a lack of interest in – or hostility to – the role of organized labour in the industry, even by many of those advocating greening the built environment.

The Adverse Effects of the Underground Economy

Canada's underground construction economy consists of a variety of practices designed to minimize labour and other costs, avoid taxes, circumvent zoning and building regulations, and, in some cases, conceal illegal activities. Its characteristics have been extensively documented in a variety of studies in recent years by both academics and government agencies (Mirus, Smith, and Karoleff 1994; Gervais 1994; Lippert and Walker 1997; Barrett 1998; Giles and Tedds 2002; Armstrong and O'Grady 2004; Expert Advisory Panel on Occupational Health and Safety 2010; Gilbert 2010; Hamilton-Smith 2010; Prism Economics & Analysis 2010; Terefe, Barber-Dueck, and Lamontagne 2012; Commission on the Reform of Ontario's Public Services 2012). Statistics Canada estimated that in 2009, underground activity in the entire Canadian economy amounted to $36 billion, of which construction accounted for 29 percent (Terefe, Barber-Dueck, and Lamontagne 2012).

A number of interrelated factors facilitate the operation of underground construction. One is the well-documented practice of tax avoidance (Auditor General of Canada 1999; Pigeon 2004; O'Grady 2010a; Terefe, Barber-Dueck, and Lamontagne 2012). Employers, small contractors, and self-employed workers avoid taxes either by not divulging that they are receiving income from those purchasing construction services or by disclosing only part of it. By not charging customers the Harmonized Sales Tax (or, in some provinces, the Goods and Services Tax), they gain a major cost advantage over legitimate contractors. They also avoid Canada Pension Plan (CPP) contributions, Employment Insurance (EI) payments, workers' compensation premiums, and other employment-related deductions (Armstrong and O'Grady 2004). Additionally, they may not declare their work to the relevant municipal authorities, thus avoiding the costs of building permits (and the oversight of building inspectors). While workers may share in some of the immediate cost savings of underground activity because they avoid paying income tax and other statutory contributions such as CPP and EI, they also lose the significant benefits that flow from these programs as well as the statutory protections associated with a formal employment relationship.

Because many of their customers – particularly in the residential construction and renovations sector – do not have the expertise to assess whether work is being done properly, underground firms can also cut corners on materials and workmanship. These practices enable them to underbid legitimate employers and contractors by a substantial margin – normally in the range of 20 to 30 percent, but in some cases as much as 50 percent (Armstrong and O'Grady 2004; Hamilton-Smith 2010). Competition from underground suppliers puts pressure on legitimate contractors to cut corners to remain competitive, including using lower-skilled or unskilled workers wherever feasible. It thus encourages deskilling. As price is normally the critical factor in determining who finds work, and assessing the quality of work is often difficult, underground operators have an enormous advantage.

While the underground economy adversely affects much of the construction industry, particularly in the residential and low-rise sectors, from the perspective of this chapter its impact on undermining green construction is of particular importance. Ruthless cost-cutting makes it difficult to implement new green technologies or use more environmentally responsible working practices because these tend to be more expensive – at least in the short run – than other options. They also normally require that those implementing these innovations have new and higher-level skills. Conflicting claims made by different contractors or manufacturers lead to confusion over the right technology, or standards, making it even more challenging for those who purchase construction services, a problem exacerbated by the absence of clear guidelines and regulations from government. Even when customers commission low-carbon innovations, they may not be properly installed because of poor workforce training, the absence of quality standards, and inadequate inspection (Gleeson and Clarke 2013).

The underground economy poses a major barrier to the development of the skills needed to implement green construction. Workers receive little or no formal training and are only rarely able to obtain recognized qualifications, such as a journeyperson's ticket (Armstrong and O'Grady 2004; Hamilton-Smith 2010). Underground employers also normally do not support apprentices. They are very small operators with little or no training capacity. They often do not have the long-term contracts needed to commit to the duration of an apprenticeship. They lack supervisory skills and do not want to incur the administrative costs of a formal training program or provide release time for classroom study. Many avoid acknowledging that they even have employees as this may raise the visibility of their activities and expose them to unwanted building inspections or tax audits. Consequently, most workers in the underground sector have no pathway to formalized training because, from the perspective of the regulatory and training authorities, they do not exist.

At the same time, as a result of their precarious job security and the fluctuating incomes that come from frequent layoffs, workers face major hurdles to investing in their own training. The result is that the large underground component of the industry fails to train its workforce. This failure clearly impacts the conventional training and apprenticeship of construction workers. But it also means that there is little opportunity for construction workers to acquire the advanced skill sets necessary for implementing low-carbon construction. While remedies for fixing the underground economy are well known – and while governments have taken some limited steps to deal with issues like tax evasion – overall, governments have failed to implement many of the obvious, urgently needed regulatory changes (Commission on the Reform of Ontario's Public Services 2012).

Aside from responding to lobbying by developers and contractors, government failure to regulate the underground economy reflects a broader commitment to neoliberal principles. To a large degree, governments accept the view that the regulatory cure – tough government measures – for the problems created by the underground economy is worse than the disease of unregulated market functioning. Despite the obvious problems, supporters of the status quo, such as the Independent Contractors and Businesses Association (ICBA) of British Columbia, maintain that there is no pressing need to regulate construction more extensively because government intervention inevitably leads to greater inefficiency and undermines economic progress.[12]

While a fundamental greening of the construction industry involves much more than simply curbing the abuses of the underground economy, this clearly needs to be a major part of the process. Government options include rigorous enforcement of the existing tax laws and comprehensive auditing of contractors, particularly self-employed owner-operators. Provincial workers' compensation boards can do more to ensure that contractors register the workers they employ and pay their assessments, while increasing the frequency of workplace inspections. Provinces can also enforce employment standards and human rights codes more vigorously (Prism Economics & Analysis 2010). And they can increase penalties for violators to a level that acts as a real deterrent, something that Quebec has done.

In Canada, only Quebec, through its Commission de la construction du Québec (CCQ), has adopted a tougher regulatory approach (Charest 2003). By providing significant funding to the CCQ, the province recovered $1.9 billion in unpaid taxes and levies during the 1997 to 2008 period. As a result of its regulatory efforts, the volume of declared work has increased much more rapidly than the growth of construction investment in the industry, indicating that more underground contractors and independent operators are complying with the law (CCQ 2014b). Quebec also requires all construction workers to be registered as a condition of working in the industry, including posting a resumé of their training and skills online. Registration also ensures that they are enrolled in the industry's pension and welfare plans.

Provinces can also license all contractors (Armstrong and O'Grady 2004). Currently, anyone can set up shop and provide construction services. However, there is no quality control, no protection for purchasers, and no guarantee that employers are following employment standards and health and safety legislation or making contributions to EI, CPP, and other programs. Licensing would force many underground operators to acknowledge that they are providing construction services and make tax and regulatory avoidance more difficult; it would also establish a database to monitor contractor performance (Hamilton-Smith 2012). Including labour standards in public procurement contracts and requiring employers to be members of employer associations, as in Quebec, would also reduce underground activity (Charest 2003). Both provincial and municipal governments can also enforce building codes more rigorously – including hiring sufficient numbers of building inspectors – to guarantee that the work being performed meets existing standards.

Some provinces, such as British Columbia, could also expand the number of compulsory trades to ensure that only those with the appropriate qualifications are able to perform key components of construction work (Armstrong 2008). This approach to labour regulation is standard practice in a number of European countries, such as Germany. More stringent licensing of the trades would also provide a major incentive for workers to take formal training or apprenticeships given that, once they completed them, the qualifications they earned would provide greater job opportunities.

To ensure that underground contractors support training financially, governments can require all firms to contribute through compulsory levies, as imposed by some Australian state governments, California, numerous European countries, and Quebec, which has a compulsory 1.5 percent-of-payroll levy (Charest 2003; Calvert 2011; Hamilton-Smith 2012). Another major policy tool is stronger employment and labour legislation. Unionized sectors of the industry have been far more successful in achieving good training outcomes than unorganized sectors. At a minimum, as in Quebec, public policy should facilitate, rather than limit, the ability of workers to unionize. This would make it much easier to implement an effective training and apprenticeship regime in the rest of Canada.

The Industry's Inadequate Training and Apprenticeship System

A second factor that needs to be addressed if the industry is to transition to low-carbon construction methods is its approach to training. Apprenticeship is the core approach to training construction workers. In most trades, approximately 80 percent of the time is spent on the job and the remainder in formal classroom training. Blocks of formal education are normally scheduled to correspond to workers' increasing on-the-job experience. Apprentices need employment to complete their training and graduate as qualified journeypersons. However, the industry experiences wide swings in employment that reflect the business cycle, seasonal factors, and the state of the local property market. In some provinces, this is exacerbated by boom-and-bust employment patterns in major infrastructure or resource projects. When the economy is booming, construction contractors are desperate for workers. When it is in recession, workers are often laid off for months or even years at a time (Sharpe and Gibson 2005; Hamilton-Smith 2010; Coe 2011). These factors lead to high labour turnover and the corresponding loss of the skills and experience that workers have acquired.

Rather than having a national, comprehensive approach to training the construction workforce, Canada has a decentralized, patchwork quilt of apprenticeship and trades training arrangements, which reflects the constitutional jurisdiction of the provinces and territories (Sharpe and Gibson 2005; Morissette 2008; Armstrong 2008; Laporte and Mueller 2010). Although the federal government provides funding through a package of training agreements with the provinces and territories, except Quebec, and several larger urban centres such as Toronto, the training system is fragmented, inadequately funded, and poorly regulated, thereby shifting most of the costs and risks to workers. Worse, instead of supporting training in low-carbon construction, the current federal government's main interest in trades training is in response to demands from the oil, gas, and mining sectors. This ignores both the need for training in green construction methods and the potential gains achievable in the built environment.[13]

The volatile nature of construction means that workers take significant risks when they commit to a three-to-five-year apprenticeship. Most have no guarantee that they will obtain steady employment, which they need to fulfill the on-the-job component of their apprenticeship program. As most are from working-class backgrounds, they normally do not have significant financial resources to carry them through prolonged periods of unemployment; this is an issue compounded by starting wage rates of approximately half that of a journeyperson. Absent adequate financial resources, workers are often forced to abandon their training to seek work elsewhere. The dropout rate in many trades is well over half of those starting an apprenticeship (Laporte and Mueller 2010; Hamilton-Smith 2010; Coe 2011).

Paradoxically, when there is ample work, apprentices may not return to school because of the loss of income and because they fear not finding work when they are finished the next phase of their classroom studies. This means losing their investment in training, while the industry loses the potential benefits of future skilled workers (Bosch and Philips 2003). These problems are inherent in an unregulated construction labour market and pose an ongoing challenge to maintaining a qualified workforce.

In practice, larger employers and/or unionized employers tend to be more supportive of training, and their apprentices have a much higher completion rate. For some employers, especially in the capital-intensive resource sector, the benefits of having a well-trained workforce are so substantial, and the costs of poorly performed work so significant, that investment in workforce skills makes economic sense. When employers are party to a collective agreement, they are also likely to support training through contract provisions stipulating the ratio of apprentices to journeypersons and negotiated contributions to training funds (O'Grady 2005, 2010a).

However, most construction employers are very small and do not conform to this pattern. Overall, the industry's training performance is very poor, with alarmingly high rates of non-completion (Gunderson 2013). Efforts by the federal and some provincial governments to encourage more workers to take apprenticeships over the past 15 years have largely failed (Hamilton-Smith 2010).

According to Coe and Herbert Emery,

> The large expansion in apprenticeship training registrations since 1991 has not produced a substantial increase in the number of persons completing apprenticeship programs. In 2007, there were only 3900 completions from apprenticeship programs in the building trades, up only 300 from 3600 completed

> apprenticeships in 1991. Moreover, the ratio of completed apprenticeships to the number of individuals registered in building trades apprenticeship programs fell from 7.7 percent in 1991 to 4.9 percent in 2007. (2012, 95)

Desjardins and Paquin of Statistics Canada used Canada's Registered Apprentice Information System to examine the completion rates of the 1994 and 1995 cohorts of starting apprenticeships ten years later. They found that completion rates for construction trades were, on average, much lower than non-construction trades. Admittedly, some building trades – electrician (62 percent), plumber (58 percent), and sheet metal worker (56 percent) – did much better than others. But trades such as steamfitter-pipefitter (38 percent) and carpenter (34 percent) did relatively poorly (Desjardins and Paquin 2010).[14]

Similarly, British Columbia's Industry Training Authority (ITA) admitted that the completion rate for apprenticeships after six years in its provincial program was only 40 percent in 2011 (ITA 2013). The ITA data does not provide a breakdown between union and non-union apprentices. But there is evidence that average completion rates would be considerably lower if the relatively high success rates of apprenticeships with unionized employers were excluded. The British Columbia and Yukon Territory Building and Construction Trades Council maintains that its union training programs have achieved a success rate averaging about 90 percent over the past five years (BCYT-BCTC 2013).

In his study of factors influencing apprenticeship completion rates, Coe (2011) noted that the rate was 10 percent higher for compulsory trades and concluded that a shift toward increasing the number of compulsory trades would likely increase this rate. Unions have been strong supporters of expanding the number of compulsory trades.

Even among larger employers, unionization is a key factor in apprenticeship. A 2013 survey by Katherine Jacobs (2013) of the Ontario Construction Secretariat found that 83 percent of unionized contractors supported apprentices, compared with only 42 percent of their non-union counterparts.

Greening the Built Environment Requires Enhanced Training and Skills

While Canadians have learned to live with the construction industry's dysfunctional approach to training and apprenticeship, global warming introduces an entirely different element into this equation. Canada now needs to green the built environment as a matter of urgency. Consequently, it needs a training system that prepares construction workers properly for this task.[15] Numerous studies indicate that low-carbon construction requires considerably higher levels of training and skills than those currently possessed by Canada's construction workforce

(Gleeson and Clarke 2013; Goodland 2012; Akenhead 2012). A recent ILO study concluded,

> Experience in a growing number of countries, both industrialized and developing, demonstrates that the construction of energy- and resource-efficient buildings requires competent enterprises and a skilled workforce. Poorly installed equipment and materials do not yield expected gains in efficiency and emissions reduction. Targeted investments in skills upgrading and certification of building firms, formalization – notably of small and medium-sized enterprises (SMEs) which dominate the sector – and improvements in working conditions to retain qualified workers are therefore essential components of a successful strategy. (ILO 2012, 127)

Successful green construction requires a highly trained workforce. Much of the work involves very tight tolerances and careful installation of components (Gleeson and Clarke 2013; Goodland 2012). It also involves learning about new technologies and new building methods. In retrofitting, where the largest energy savings are possible, the skilled trades must be able to handle a wide variety of different building types, erected at different times using different technologies and different materials. To do this work well, trades must have a comprehensive knowledge of many different building systems and technologies so that they are able to apply the most effective approach to each specific project. This requires the development of higher skill levels and a significant, ongoing investment in workforce training.

Decisions about what needs to be done to green the existing building stock require both an understanding of the particular attributes of individual buildings and a knowledge of the options now available to reduce energy consumption. These options must also be assessed in light of their cost-effectiveness, installation complexity, skill requirements and skills availability, and a variety of other factors. They require an understanding of the fundamentals of building construction and the possibilities arising from the new green technologies, materials, and systems now available. This requires knowledge, judgment, and a clear understanding of the challenges that have to be overcome.

Construction workers are constantly involved in problem-solving on the work site. They normally do this without close supervision. Often, on smaller projects, only one or two members of a trade are responsible for handling the work, and they are expected to do it on their own in a way that meets the professional standards mandated by the architects, engineers, or prime contractors. While there are tasks on building sites that can be performed by workers with little or no training or skills, retrofitting does not readily fit into this pattern. In reality, every project is different, and every project has its unique challenges, many of which are not predictable. Consequently, the trades have to be able to analyze the characteristics of a wide variety of building projects and make judgments about how best to implement the objectives of these projects in a cost-effective, safe, and efficient way.

Similarly, establishing more stringent energy conservation requirements in building codes is not only contingent on having an effective enforcement system to ensure that builders comply. It also requires a workforce that knows the code requirements and is capable of implementing them on the building site. Detailed knowledge of building codes requires a training system that enables workers to acquire this information and to continually update it as green protocols are introduced. Continuing reliance on a largely unskilled construction workforce is not compatible with achieving the objectives of low-carbon construction.

The Lack of Involvement of the Skilled Trades and Their Unions in Implementing Green Construction

A third barrier to the greening of the construction industry is the limited role that building trades workers and their unions currently play in this process. Outside Quebec, less than a third of the industry is organized; union representation is concentrated in the ICI sector and primarily with large contractors. It is also uneven, varying considerably among the trades and between the trades and semi-skilled or unskilled construction workers. Although average union density in construction has remained fairly constant at just under a third of the workforce during the past 15 years in Canada, after falling significantly during the 1980s and early 1990s, this figure conceals major shifts in the pattern of unionization, with increases in Quebec offsetting declines in English Canada (Charest 2003; Galarneau and Sohn 2013; Statistics Canada 2014a).

TABLE 7.1
Construction Collective Agreement Coverage in Canada 1997–2012 (% of workforce)

Province	*1997*	*2000*	*2003*	*2006*	*2009*	*2012*
Newfoundland and Labrador	25.71	27.27	20.78	27.10	26.62	31.95
Prince Edward Island	20.59	13.79	19.35	13.95	17.95	23.68
Nova Scotia	38.16	26.97	22.78	26.40	23.98	28.99
New Brunswick	27.78	29.41	20.00	21.74	29.72	25.13
Quebec	48.48	50.11	56.25	57.69	57.80	59.30
Ontario	32.62	32.42	32.99	29.52	30.51	31.51
Manitoba	22.56	21.47	17.34	27.83	22.48	22.22
Saskatchewan	22.86	24.66	23.31	22.01	17.67	21.60
Alberta	17.19	21.04	21.67	18.78	19.29	19.38
British Columbia	31.31	31.77	30.51	25.02	20.35	21.89
All-Canada union coverage	32.39	32.53	34.19	31.67	31.23	32.76

Source: Statistics Canada (2014a).

Overall, the influence of the building trades unions – like the larger Canadian labour movement – has been on the decline. Despite this setback, some unionized trades continue to play a major role in establishing and enforcing their qualifications; such is the case with the International Brotherhood of Electrical Workers (IBEW) through its role in the National Electrical Trade Council (NETCO) (MacLeod 2012). Construction unions in Ontario have persuaded the provincial government to give them a role as members of the board overseeing the construction training and licensing component of the newly established Ontario College of Trades. And building trades unions in certain regions, such as the Greater Toronto Area, continue to organize significant numbers of workers in major commercial and industrial projects (although the residential sector remains largely unorganized).

Many unions continue to play a significant role in training and apprenticeship. In Ontario, for example, they operate over 200 trades training facilities in co-operation with unionized employers (O'Grady 2005). The building trades unions have collective agreement provisions requiring their employers to contribute part of the wage package – normally about 1 percent of payroll (in the form of a specified number of cents per hour) – to Training Trust Funds (TTFs). TTFs are jointly managed by a union and a group of contributing employers with which the union has collective agreements. They provide part, or all, of the classroom training that apprentices require, in some cases supplemented by provincial government purchases of training seats. They may also share some of the classroom training with publicly funded community colleges or purchase classroom capacity from them.

In an industry characterized by numerous small employers, few of which have the capacity to guarantee apprenticeships themselves or provide the classroom component of training, TTFs fulfill a vital industry need. Most apprentices work for a number of employers during their training, and their union affiliation provides employment continuity during this period. Because they are based on multi-employer collective agreements, TTFs tend to have stable funding, which facilitates long-term planning of courses and programs. In Ontario, they provide some training to 25 percent of building union members each year (O'Grady 2005). They also have an excellent track record of facilitating a high percentage of apprentices to actually complete their apprenticeships – a rate far higher than in the non-union sector. Union training accomplishments contrast sharply with the failure of various federal and provincial government policy initiatives, such as providing training tax credits to employers (Hamilton-Smith 2012).

In Quebec, the building trades have significantly expanded their role in training and apprenticeship, a development that reflects the provincial government's very different approach to construction labour relations and its willingness to require employers to take more responsibility for training through its payroll levy system. The provincial government's 1968 legislation – an *Act Respecting Labour Relations, Vocational Training and Manpower Management in the Construction Industry* – has encouraged collective bargaining in the construction sector and resulted in relatively high union density (Quebec 1968).[16] Quebec unions have been able to establish a wide network of TTFs through negotiations with unionized employers, and these continue to play a central role in trades training in the province (Charest 2003; O'Grady 2005). According to the Commission de la Construction du Quebec, its $150 million training fund supports 20,000 trainees per year for apprenticeship and continuing trades training (CCQ 2014a).

Unions have also supported training through project agreements such as British Columbia's Vancouver Island Highway Project. This was the largest construction project in western Canada during the mid-1990s. The provincial government created a public corporation to act as employer of all construction labour on the project. It negotiated a blanket agreement with 13 highway construction unions, which included commitments to local hiring, local training, and employment for members of the province's four equity groups. Union training facilities played an important role in this process, as did the commitment of union officials to provide technical and staff support for the program. In an industry in which equity group participation had been about 2 percent overall, the Vancouver Island Highway Project achieved equity participation of just over 20 percent of all hours worked during its eight years (Cohen and Braid 2000; Calvert and Redlin 2003).

That there is significant potential for unions to play a greater role in climate initiatives is evidenced by a number of other promising initiatives. One is the work of British Columbia's International Association of Heat and Frost Insulators and Allied Workers Local 118 (BC Insulators). Its members insulate water pipes, heating installations, storage tanks, and air conditioning and refrigeration systems. Normally referred to as mechanical insulation (MI), this building component is key to reducing energy use by minimizing heat and refrigeration losses. Done properly, it can often lower energy consumption in buildings by as much as half.

BC Insulators have strongly supported the development of new green building standards. But in recent years, they noted that the LEED gold, platinum, and other requirements specified by developers, architects, and engineers were not being met because of faulty workmanship by contractors and unskilled installers. They also found that too many projects did not have an integrated, comprehensive approach to energy conservation because they ignored the importance of proper MI. For example, mechanical steam piping was often being installed behind drywall with no insulation and no access to remedy the resulting problem, causing mould, respiratory hazards, and massive loss of energy.

As a result, the union commissioned a study to document the extent of energy losses and related problems commonly found in newly constructed buildings, including many that had a green designation (HB Lanarc Consultants 2010). Based on this study, the union developed a comprehensive package of new building code measures and government procurement requirements explicitly designed to reduce GHG emissions and energy consumption. The union's business manager, Lee Loftus, and staff representatives have made presentations to every major municipality in British Columbia and promoted the union's approach both nationally and at local-government conferences across the United States. The union has also lobbied the BC government for much tougher building code standards and for new provisions in public procurement that would require improved energy efficiency in public buildings (HB Lanarc Consultants 2010). According to Loftus, its 365 members have spent over $75,000 on their campaigns over the past five years (Interview, 7 June 2012).

On the training side, BC Insulators have worked closely with the British Columbia Institute of Technology to develop new climate modules to include in the classroom training program for apprentices in the insulating trade. They have also lobbied for a new program to train inspectors in MI. And they have worked extensively with both the environment committee of the BC Federation of Labour and Green Jobs BC in support of tougher climate policies.

Similarly, at the national level, the IBEW has worked closely with industry through NETCO to develop new standards and training modules for the installation of solar photovoltaic (PV) equipment and related renewable electric-vehicle technology. The union has played a central role in developing the new standards and related training modules. Qualified electricians can now add a certification in these areas to their trade qualifications (NETCO 2011).[17]

The implementation of low-carbon construction on the job site aligns well with the building trades unions' goal of maintaining high standards of skills and supporting the apprenticeship and training system (Lowe 2009).[18] True, their goal of a high-skill, high-wage industry reflects their self-interest in maintaining or improving the wages of their members as well as a desire to exercise control over the labour process. Nevertheless, most skilled trades take pride in exercising their skills and being able to adapt to a wide range of problem-solving situations – a key feature of effective climate retrofitting.

The interests of building trades unions also correspond to the need to have a more stable and permanent workforce with a long-term attachment to the industry. Their members want a career, not just a job. This, too, is a precondition for expanding and maintaining the knowledge base essential for implementing low-carbon construction. High turnover in the industry, especially among trainees who fail to complete their apprenticeships, is not conducive to effective climate programs. Construction unions also support a workplace culture that values work well done and takes pride in learning and exercising demanding skills (Sennett 2009). The challenge of incorporating climate issues into a well-trained, stable workplace that already supports these values is far less than one that attempts to do so in a workforce that is poorly skilled, precarious, and low paid and that has little or no long-term attachment to the industry.

Despite the evidence that apprentices in unionized workplaces have a high completion rate and that unions have consistently advocated maintaining a high-quality training system, most policy-makers do not see unions as major players in the green transformation of the industry. In much of the literature on greening construction, unions simply do not exist. Some provincial governments, such as British Columbia's, have excluded unions from provincial training authorities, rewriting labour codes, and procurement policies in order to marginalize organized labour and, not insignificantly, to deny construction workers a voice in shaping industry training.[19]

Similarly, outside the specific organizations that unions and unionized employers have jointly created to promote training and workplace skills, such as TTFs in Ontario and Quebec and NETCO and its national counterparts in other trades, government policy-makers and many advocates of greening the built environment generally ignore the potential contribution of unions. The assumption is that low-carbon construction will be introduced successfully without the active involvement of workers – other than by doing what they are assigned to do.[20]

One reason is the general shift in Canada toward a neoliberal labour relations paradigm, in which the role of all unions is marginalized. The labour climate has become increasingly hostile to unions at both the federal and the provincial levels, as evidenced by regressive changes to labour legislation and the increasing tendency of governments to impose back-to-work legislation whenever unions strike. In the context of this broader anti-labour shift, arguments about the potential benefits of a greater union role in training, apprenticeship, and climate change initiatives go against the tide. Yet the evidence is clear: a highly skilled construction workforce is now needed to implement climate objectives, and unions have demonstrated that they can contribute to this process.

CONCLUSION

This chapter has identified three major problems that are impeding the ability of the Canadian construction industry to meet the challenge of global warming: the extensive and largely unregulated underground economy, fundamental weaknesses in the current training and apprenticeship system, and the failure to recognize the potential contribution of building trades workers – and the unions that represent them – in the process of greening the built environment. It does not claim that these are the only, or the most important, impediments.[21] But it does assert that addressing these three problems is an essential part of what needs to be done.

Effective climate action requires a comprehensive approach to reducing energy consumption and GHG emissions in the built environment. Given the size of the existing stock of buildings, and the urgency of reducing GHG emissions and energy use, there is an enormous potential to create new construction jobs in energy conservation and retrofitting.[22] The development of a workforce with suitable skills and training in advanced low-carbon construction techniques is an essential part of this process.

Unfortunately, the industry's current priorities – and particularly its approach to workforce training and development – are major barriers to greening the industry. A commitment to unregulated market principles, hostility to public regulation, failure to tackle the abuses of the underground economy, reluctance to resource training and apprenticeship adequately, and opposition to a larger role for unions are important barriers to achieving a workforce capable of implementing low-carbon construction. Absent significant changes to the basic organization of the construction industry, it will continue to lack the capacity to implement the climate measures in the built environment that are now so urgently needed.[23]

NOTES

1. Some of the background research included in this chapter formed the basis for an earlier paper published in *Alternative Routes* (Calvert 2014).
2. The CSC recently changed its name to BuildForce Canada; however, the papers on its website still use the earlier name and are cited accordingly.
3. In the context of this chapter, we are defining the built environment primarily in terms of the various structures and their internal systems that use energy and contribute to global warming. However, some definitions of the built environment are much broader, encompassing the impact of neighbourhood design, transportation infrastructure, and manufactured facilities such as water and sewer systems that use energy and produce GHG emissions.
4. This is not to suggest that improving the energy efficiency of work performed in the resource sector should be ignored; clearly, there are gains to be made. However, the built environment has much more potential and, therefore, is by far the highest priority.
5. The terminology for describing general contractors varies somewhat in the industry. Another term commonly used is prime contractor.
6. This chapter focuses on construction in English Canada. As it notes in various places, the situation in Quebec differs quite substantially from that in the rest of Canada. Many of the problems identified in this chapter are far less pronounced in that province because public regulation of the industry is much more extensive. Charest (2003) provides a good analysis of the Quebec industry.

7. This is compounded by the fact that most purchasers of construction services are not knowledgeable about the details of the work and are not in a position to evaluate matters such as quality of design, materials, installation methods, or workmanship. Too often, they are not aware of the options available to reduce the carbon footprint of their buildings or the potential long-term financial benefits of reduced energy use. The classic principal-agent dilemma is characteristic of much of the industry's operation, reinforced by asymmetry of information between purchaser and contractor (O'Grady 2010b; Hamilton-Smith 2012; Goodland 2012). Additionally, when landlords do not pay the cost of heat and electricity, they often see little benefit in paying for improved energy conservation in their buildings, while tenants have no reason – or capacity – to invest in improved insulation or other energy conservation initiatives in buildings that they do not own and in which they may not reside long enough to reap the benefits (Levine and Ürge-Vorsatz 2007; UNEP 2009).
8. One of the most egregious examples of this is in the demolition of existing buildings. Because it is normally far cheaper to bulldoze an old structure than to dismantle it so that its materials can be reused, massive volumes of used materials end up in landfill sites. The industry prefers this option as it minimizes labour expenditures, which are the major cost in taking apart a building. Industry influence has largely undermined efforts by municipal governments to implement effective materials recycling as this would require extensive public regulation of demolition practices and the creation of new markets for used materials.
9. In recent years, there has been a great deal of interest in how global supply chains now involve extensive subcontracting of much of the world's manufacturing to developing countries that have low wages, minimal taxes, and little worker protection. But, in reality, this pattern has been part of the construction industry in Canada for generations. Subcontracting is the norm, with consequences that parallel what has been happening more recently in the extensive outsourcing by corporations in the developed world to subcontractors in the developing world.
10. BuildForce Canada has slightly different figures, although the basic pattern is similar: 90 percent of residential construction firms and 70 percent of non-residential firms have fewer than five workers. Only 1 percent of residential firms employ more than 50.
11. The demarcation lines between various trades are, arguably, also a contributor to industry fragmentation as they reinforce a proprietary approach to work in which control over skill sets (and the unions representing workers with these skill sets) results in occupational silos. This can impede broader worker solidarity as well as more effective integration of all work on a building site, particularly as there are significant differences in wages among various trades. In countries such as Denmark, where the wage structure is very flat, unions have been able to overcome these divisions more effectively than in Canada as demarcation issues do not significantly impact wages.
12. See, for example, the anti-government focus of the ICBA's publications for the most extreme version of this view.

13. While provinces have made significant efforts to establish national standards that are recognized across the country through the Red Seal program, there are still significant variations in provincial and territorial approaches. This is highlighted by the difference in the number of designated compulsory trades in which practitioners must have a certificate of trades qualification as a legal condition of performing a particular type of construction work. While some provinces have several dozen such trades, others have as few as three.
14. While Statistics Canada (2014c) data for 2011 indicates an increase in completions to 55,422, the number of apprentices registered in the system is also up, to 426,283.
15. This is not to imply that the industry will automatically provide jobs to workers who have acquired green skills. Absent much tougher government requirements that force industry to adopt low-carbon construction, the economics of the labour market will determine employment. However, if there are fundamental changes in the industry's support for green building techniques, then a qualified workforce will be essential.
16. The act has undergone some revisions in the intervening years. For example, in the title, the word *Manpower* has been replaced by *Workforce*. In the fall of 2013, an amendment brought in a regulation to establish the *Carnet référence construction*, a mandatory online referral service for the construction industry. Construction workers post a resumé outlining their skills and training on the Carnet's website, while employers use the site as a clearing house for finding workers (http://carnet.ccq.org/en/).
17. Carol MacLeod (NETCO executive director), interview with the author, 12 August 2013; and Andy Cleven (director of training, Electrical Joint Training Committee, Electrical Contractors Association of British Columbia), interview with the author, 12 July 2012.
18. Lee Loftus, interview with the author, 7 June 2012; Cleven, ibid.
19. The BC government's recent interest in discussing training issues with the province's building trades unions may appear to signal a change in focus under Christie Clark. However, its efforts have focused on fulfilling labour shortages in the expanding energy sector as part of the province's economic agenda of supporting energy and mining development. Retrofitting the built environment is significantly absent from its policy focus.
20. The federal government's trades training initiatives have focused on providing skills for its favoured resource extraction industries, where it needs the training expertise of a number of the construction unions. The goal is to encourage unions to address industry needs in the absence of other apprenticeship and training options that the federal government might otherwise prefer. It is arguably courting some of the major building trades unions not because it supports unionization but because it – and resource industry employers – needs them, at least for the immediate future. However, its policies are designed to enable employers, not unions or workers, to have a greater say about where public money for training will be allocated. For example, its controversial Canada Job Grant program provides no new money. Rather, it will shift $300 million per year from existing Canada Labour Market Agreement training programs to its more employer-focused approach. The federal government will provide employers with up to $5,000

per trainee, to be matched by an equal contribution by the provinces and, in theory, employers. However, these initiatives to meet the needs of employers have not been accompanied by any policies designed to expand union representation. Significantly, there is virtually no mention of unions in the various employment and training initiatives outlined in Canada's Economic Action Plan 2013 except for its Helmets to Hardhats program for veterans (Canada 2013).

21. Thus, we have not discussed issues such as the major role that publicly funded trades training colleges can play, a greater emphasis on climate issues in the education of architects and engineers, the comprehensive greening of municipal planning and building codes, and direct government financial support for training and apprenticeships, to cite a few of the more obvious areas where major climate initiatives are also needed.
22. An example of this is found in the approach that the Danish construction unions took to the economic crisis of 2008. Recognizing that a stimulus package was needed to address the crisis, they lobbied the government (successfully) to bring forward a number of major building retrofitting programs, which they argued would both create jobs and reduce the country's carbon footprint (Calvert 2011).
23. In addition to the economic and environmental arguments in support of major change, there is a strong moral case that workers, like other citizens in a democratic society, have the right to a voice in shaping the development of our future climate policies, whether in construction or in other parts of the Canadian economy. Policy should recognize this right (Regan 2010).

REFERENCES

Akenhead, S. 2012. *Report from the 2nd Greening the Trades Consultation, Ladysmith, BC*. Commissioned by the British Columbia Institute of Technology.

Armstrong, T.E. 2008. *Compulsory Certification Project*. Toronto: Ministry of Training, Colleges and Universities.

Armstrong, T.E., and J. O'Grady. 2004. *Attacking the Underground Economy in the ICI Sector of Ontario's Construction Industry*. Toronto: Ontario Construction Secretariat. http://www.iciconstruction.com/admin/contentEngine/dspDocumentDownload.cfm?PCVID=ddcf803a-1372-0ea9-42b5-12e2ff58d034.

Auditor General of Canada. 1999. *Report of the Auditor General of Canada: Chapter 2 – Revenue Canada – Underground Economy Initiative*. Ottawa: Office of the Auditor General. http://www.oag-bvg.gc.ca/internet/docs/9902ce.pdf.

Barrett, D. 1998. "The Renewal of Trust in Residential Construction." Victoria: Commission of Inquiry into the Quality of Condominium Construction in British Columbia. http://www.qp.gov.bc.ca/condo/.

BCYT-BCTC (British Columbia and Yukon Territory Building and Construction Trades Council). 2013. "Highlights from 2012 Apprenticeship Survey." Burnaby: BCYT-BCTC.

Bosch, G., and P. Philips, eds. 2003. *Building Chaos: An International Comparison of Deregulation in the Construction Industry*. London: Routledge.

Calvert, J. 2011. "Climate Change, Construction and Labour in Europe: A Study of the Contribution of Building Workers and Their Unions to 'Greening' the Built Environment in Germany, the United Kingdom and Denmark." Working

paper presented at "Greening Work in a Chilly Climate," the Work in a Warming World (W3) Researchers' Workshop, Toronto, November 2011.
—. 2014. "Overcoming Systemic Barriers to 'Greening' the Construction Industry." *Alternative Routes: A Journal of Critical Social Research* 25:81–118.
Calvert, J., and B. Redlin. 2003. "Achieving Public Policy Objectives through Collective Agreements: The Project Agreement Model for Public Construction in British Columbia's Transportation Sector." *Just Labour* 2:1–13.
Canada. 2013. "Canada's Economic Action Plan: Canada Job Grant." http://actionplan.gc.ca/en/initiative/canada-job-grant.
CCQ (Commission de la construction du Québec). 2014a. "Training." http://www.ccq.org/en/DevenirTravailleur/F_Formation/.
—. 2014b. "Workforce Changes." http://www.ccq.org/en/Patronales/J_ResponsabilitesEmployeur/J03_MouvementsMainDoeuvre.
Charest, J. 2003. "Canada Labour Market Regulation and Labour Relations in the Construction Industry: The Special Case of Quebec in the Canadian Context." In *Building Chaos: An International Comparison of Deregulation in the Construction Industry*, ed. G. Bosh and P. Philips, 95–136. London: Routledge.
Coe, P.J. 2011. "Apprenticeship Program Requirements and Apprenticeship Completion Rates in Canada." Working Paper 71. Ottawa: Canadian Labour Market and Skills Researcher Network. http://www.clsrn.econ.ubc.ca/workingpapers/CLSRN%20Working%20Paper%20no.%2071%20-%20Patrick%20Coe.pdf.
Coe, P.J., and J.C. Herbert Emery. 2012. "Accreditation Requirements and the Speed of Labour Market Adjustment in Canadian Building Trades." *Canadian Public Policy* 38 (1):91–111.
Cohen, M.G., and K. Braid. 2000. "Training and Equity Initiatives on the British Columbia Vancouver Island Highway Project: A Model for Large-Scale Construction Projects." *Labor Studies Journal* 25 (3):70–103.
Commission on the Reform of Ontario's Public Services. 2012. *Public Services for Ontarians: A Path to Sustainability and Excellence*. Toronto: Queen's Printer for Ontario. http://www.fin.gov.on.ca/en/reformcommission/chapters/report.pdf.
CSC (Construction Sector Council). 2011. "Green, Sustainable Building in Canada: Implications for the Commercial and Residential Construction Workforce – G2011." Ottawa: CSC.
—. 2013. "Construction Looking Forward: 2013–2021 Key Highlights – National Summary." Ottawa: CSC.
Desjardins, L., and N. Paquin. 2010. *Registered Apprentices: The Cohorts of 1994 and 1995, One Decade Later*. Research Paper. Cat. no. 81-595-M – No. 080. Ottawa: Statistics Canada.
Expert Advisory Panel on Occupational Health and Safety. 2010. *Report and Recommendations to the Minister of Labour*. Toronto: Ministry of Labour.
Galarneau, D., and T. Sohn. 2013. "Long-Term Trends in Unionization." Article: Insights on Canadian Society. Cat. no. 75-006-X. Ottawa: Statistics Canada.
Gervais, G. 1994. "The Size of the Underground Economy in Canada." Studies in National Accounting, no 2. Cat. no. 13-603E, No. 26. Ottawa: Statistics Canada.
Gilbert, R. 2010. "Harmonized Sales Tax Could Drive Contractors to Underground Economy." *Journal of Commerce* 53, 5 July. http://www.journalofcommerce.com/Home/News/2010/7/Harmonized-sales-tax-could-drive-contractors-to-underground-economy-JOC039537W/.

Giles, D., and L. Tedds. 2002. "Taxes and the Canadian Underground Economy." Canadian Tax Paper No. 106. Toronto: Canadian Tax Foundation.

Gleeson, C., and L. Clarke. 2013. "The Neglected Role of Labour in Low Energy Construction: 'Thermal Literacy' and the Difference between Design Intention and Performance." Paper presented at the W3 International Conference 2013, Toronto, 29 November–1 December.

Goodland, H. 2012. "BCIT: Greening the Trades." Rev. 4. Victoria: Brantwood Consulting.

Gunderson, M. 2013. "Completion Counts: Raising Apprenticeship Completion Rates in Ontario's Construction Industry." Toronto: Ontario Construction Secretariat. http://www.iciconstruction.com/admin/contentEngine/contentDocuments/Completion_Counts_-_Final_Report_-_OCS_2013_-_E-Version.pdf.

Hamilton-Smith, E. 2010. "Deterring the Underground Economy in BC's Residential Construction Sector Demand-Side Policy Directions." MPP Paper 827 – G100. Burnaby, BC: Simon Fraser University.

—. 2012. "Increasing Skilled Trades Employer Participation in Apprenticeship Training in British Columbia." Master's thesis, School of Public Policy, Simon Fraser University, Burnaby, BC.

HB Lanarc Consultants. 2010. "Pipes Need Jackets Too: Improving Performance of BC Buildings through Mechanical Insulation Practice and Standards – A White Paper." Vancouver: International Association of Heat and Frost Insulators and Allied Workers – Local 118. http://www.energyconservationspecialists.org/wordpress/wp-content/uploads/2011/02/White-Paper_Pipes-need-jackets-too.pdf.

ILO (International Labour Organization). 2012. *Working Towards Sustainable Development: Opportunities for Decent Work and Social Inclusion in a Green Economy*. Geneva: ILO. http://www.ilo.org/wcmsp5/groups/public/---dgreports/---dcomm/---publ/documents/publication/wcms_181836.pdf.

ITA (Industry Training Authority). 2013. *Your Ticket to Trade Skills: Annual Service Plan Report 2012/2013*. Richmond, BC: ITA. http://www.itabc.ca/sites/default/files/docs/about-ita/corporate-reports/ITA_AR12-13_Final.pdf.

Jacobs, K. 2013. "Ontario's Construction Outlook: Optimism Strengthens." Presentation at the 13th Annual Ontario Construction Secretariat State of the Industry & Outlook Conference 2013, Toronto, 6 March. http://www.iciconstruction.com/admin/contentEngine/dspDocumentDownload.cfm?PCVID=45f02f53-ece7-a3dd-71a9-c7e194203582.

Laporte, C., and R. Mueller. 2010. "The Persistence Behaviour of Registered Apprentices: Who Continues, Quits, or Completes Programs?" Canadian Labour Market and Skills Researcher Network Working Paper 62. Vancouver: Department of Economics, University of British Columbia.

Levine, M., and D. Ürge-Vorsatz. With K. Blok, L. Geng, D. Harvey, S. Lang, G. Levermore, et al. 2007. "Residential and Commercial Buildings." Chapter 6 of *Climate Change 2007: Mitigation – Contribution of Working Group III to the Fourth Assessment Report of the Intergovernmental Panel on Climate Change*. Cambridge: Cambridge University Press. www.ipcc.ch/publications_and_data/ar4/wg3/en/ch6.html.

Lippert, O., and M. Walker, eds. 1997. "The Underground Economy: Global Evidence of Its Size and Impact." Vancouver: Fraser Institute.

Lowe, G.S. 2009. "People and Performance: Building Alberta's Future Construction Workforce." Discussion Paper. Calgary: Building Trades of Alberta.

MacLeod, C. 2012. "Red Seal Electrical Trades in Canada: A Policy Framework on National Standards, Apprenticeship and Journeyperson Skills Training." Ottawa: National Electrical Trade Council. www.ceca.org/netco/NETCO%202012%20Policy%20Document.pdf.

Mirus, R., R.S. Smith, and V. Karoleff. 1994. "Canada's Underground Economy Revisited: Update and Critique." *Canadian Public* Policy XX (3):235–52.

Morissette, D. 2008. *Registered Apprentices: The Cohort of 1993, a Decade Later, Comparisons with the 1992 Cohort*. Cat. no. 81-004-X. Ottawa: Statistics Canada.

NETCO (National Electrical Trade Council). 2011. *National Occupational Analysis: Construction Electrician–Solar Photovoltaic (PV) Systems Certification*. Summary Report. Toronto: NETCO.

O'Grady, J. 2005. *Training Trust Funds: A Review of Their History, Legal Foundations, and Implications for Trade Union Training Strategy*. Final Report. Toronto: Canadian Labour Congress. www.ogrady.on.ca/Downloads/Papers/Training%20Trust%20Funds.pdf.

—. 2010a. *Estimates of Revenue Losses to Governments as a Result of Underground Practices in the Ontario Construction Industry: 1995–1997 Compared to 1998–2000; Updated Estimates – August 2001*. Toronto: Ontario Construction Secretariat.

—. 2010b. "The Impact of Climate Change on Employment and Skills Requirements in the Construction Industry." In *What Do We Know? What Do We Need to Know? The State of Research on Work, Employment and Climate Change in Canada*, ed. Carla Lipsig-Mummé, 167–94. Toronto: Work in a Warming World Research Programme. http://warming.apps01.yorku.ca/wp-content/uploads/2011/08/What-do-we-know-full-report-final.pdf.

Pigeon, M.-A. 2004. *The Underground Economy: Measurements and Consequences*. PRB-04-40E. Ottawa: Parliamentary Information and Research Service, Library of Parliament.

Prism Economics & Analysis. 2010. *Underground Economy in Construction: It Costs Us All*. Toronto: Ontario Construction Secretariat. http://www.iciconstruction.com/admin/contentEngine/contentDocuments/2010_OCS_Underground_Economy_Report_FULL_BOOK_for_web.pdf.

Quebec. 1968. *An Act Respecting Workforce Vocational Training and Qualification*. CQLR c F-5. http://canlii.ca/t/kncb.

Regan, K.H. 2010. "The Case for Enhancing Climate Change Negotiations with a Labor Rights Perspective." *Columbia Journal of Environmental Law* 35 (1):249–84.

Sennett, R. 2009. *The Craftsman*. London: Penguin Books.

Sharpe, A., and J. Gibson. 2005. *The Apprenticeship System in Canada: Trends and Issues*. CSLS Research Report 2005-04. Ottawa: Centre for the Study of Living Standards.

Statistics Canada. 2013. "Gross Domestic Product (GDP) at Basic Prices, by North American Industry Classification System (NAICS): Monthly. CANSIM Table 379-0031. http://www5.statcan.gc.ca/cansim/a26?lang=eng&retrLang=eng&id=3790031&paSer=&pattern=&stByVal=1&p1=1&p2=31&tabMode=dataTable&csid=.

—. 2014a. "Labour Force Survey Estimates (LFS), Employees by Union Coverage, North American Industry Classification System (NAICS), Sex and Age Group, Annual." CANSIM Table 282-0078. Last modified 10 January. http://www5.

statcan.gc.ca/cansim/a05?lang=eng&id=2820078&pattern=2820078&searchTypeByValue=1&p2=35.

—. 2014b. "Labour Force Survey Estimates (LFS), Employment by North American Industry Classification System (NAICS), Seasonally Adjusted and Unadjusted, Monthly." CANSIM Table 282-0088. Last modified 10 October. http://www5.statcan.gc.ca/cansim/a05?lang=eng&id=2820088.

—. 2014c. "Registered Apprenticeship Training, Completions, by Age Groups, Sex and Major Trades Groups, Annual." CANSIM Table 477-0054. Last modified 11 June. http://www5.statcan.gc.ca/cansim/a26?lang=eng&retrLang=eng&id=4770054&paSer=&pattern=&stByVal=1&p1=1&p2=-1&tabMode=dataTable&csid=.

Terefe, B., C. Barber-Dueck, and M. Lamontagne. 2012. "Revisions to Its 2011 Study: Estimating the Underground Economy in Canada 1992–2008. Ottawa: Statistics Canada, Income and Expenditure Accounts Division." June. http://www.cra-arc.gc.ca/nwsrm/fctshts/2012/m09/fs120927-eng.html.

UNEP (United Nations Environment Programme). 2009. *Buildings and Climate Change: Summary for Decision-Makers*. Paris: UNEP Sustainable Buildings and Climate Initiative. http://www.unep.org/SBCI/pdfs/SBCI-BCCSummary.pdf.

CHAPTER 8

LABOUR AND THE GREENING OF HOSPITALITY: RAISING STANDARDS OR UNION GREENWASHING?

Steven Tufts and Simon Milne

INTRODUCTION

In recent years, a number of labour union strategic initiatives have been developed that seek to leverage consumer preferences against employers in the hospitality sector. These programs include rating and certifying hotels based on environmentally and socially responsible behaviour and worker-friendly practices as well as advocating for labour and environmentally sustainable practices in food services. The union campaigns are a response to management's perceived "greenwashing"[1] of hospitality services through spurious claims of local sourcing and voluntary, self-reporting rating systems. This chapter examines two union campaigns that attempt to integrate social and environmental best practices: the First Star program (Australia) and the Real Food Real Jobs campaign (United States and Canada). We find that environmental criteria are integrated into these campaigns, but they remain subservient to labour issues. These campaigns have made raising standards in the workplace a higher priority than committing to environmental sustainability, and this, plus their limited geographical scale, will likely challenge their future success.

This chapter develops an understanding of the above limitations by applying a labour geography perspective, which focuses on how workers and their institutions (e.g., labour unions) exercise power in capitalist economic landscapes (Herod 1998, 2001; Castree 2007). Labour geographers

Work in a Warming World, edited by Carla Lipsig-Mummé and Stephen McBride. Kingston: School of Policy Studies, Queen's University.

have recently begun to look at how nature intervenes in capital-labour relations, to the point where we can speak of a capital-labour-nature trialectic (Carey and Tufts 2014; Nugent 2011). Inspired loosely by the approach taken by Wainwright and Mann (2012) to conceptualize different governance scenarios as states address climate change on a global scale, we develop a matrix of capital-labour-nature relations. We believe that labour geography can take more seriously its theorizations in the context of the current Anthropocene.

We begin with a conceptual discussion, followed by a brief review of greenwashing in the hospitality sector (with a focus on accommodation services). The chapter then turns to the two cases of union campaigns that integrate social, economic, and environmental issues. The research draws on web-based material, union documents, and selected interviews with union staff about the genesis of the campaigns. We conclude with a discussion of the limits and possibilities of these strategies for hospitality workers.

LABOUR GEOGRAPHY IN THE ANTHROPOCENE

Labour geography emerged as a field of study largely following an intervention in the 1990s by economic geographers, who identified an imbalance in theoretical approaches to explaining changing economic landscapes, which gave precedence to the power of capital over that of labour (Herod 1997, 1998). Recent commentaries have evaluated the labour geography project, tracing its intellectual development and identifying key themes, areas of debate, and future directions (Lier 2007; Castree 2007; Tufts and Savage 2009; Herod 2010; Rutherford 2010; Coe and Jordhus-Lier 2011). First and foremost is concern over how to best conceptualize the agency of labour in contemporary economic landscapes. Second is the question of how we theorize the geographical scale of labour action. Third, there are significant criticisms of how labour is often conflated with class as one conceptual category (Das 2012). Last, there are still issues about how labour geography theorizes the rôle of states and how workers shape their own political regulation (Rutherford 2013; Jordhus-Lier 2012).

Recent work has also attempted to integrate socio-ecological questions into a geographical understanding of capital-labour relations (Prudham 2005; Nugent 2011). Integrating questions concerning the production of nature into a labour geography perspective adds layers of complexity but is unavoidable in the Anthropocene, a new geological period where the earth's physical processes and conditions are being significantly altered by human action (Crutzen and Stoermer 2000). Labour is in a dialectical relationship with capital, but nature inserts itself in ways that can no longer be economically, politically, socially, culturally, and theoretically ignored. As argued elsewhere, labour geography is an approach that can explore how the production of nature intersects with labour's and capital's production of economic landscapes in a capitalist system (Carey and Tufts 2014).

At a very basic level, questions arise over how much real power labour can exercise in shaping economic landscapes when nature imposes serious limits on accumulation. Eco-socialists have argued that Marx himself was more than aware of the challenges that natural limits pose for capital accumulation (Burkett 1999, 2006). At the same time, capitalism has demonstrated significant capacity to withstand both economic and environmental crises. Considering the agency of labour in this process and its role in the production of nature (Smith 1984) was prescient.

Labour geographers have focused a great deal on the scale of organization (Herod 2011). Indeed, working-class solidarity at a global scale and the geographical dilemma that confronts workers in unevenly developed economies (Castree et al. 2004) is perhaps the central problem of labour geography. How workers scale up action to confront global flows of capital, which more easily achieves spatial fixes (e.g., relocating to areas with cheap labour and fewer environmental regulations), drives inquiry. Labour's production of a global scale of action seems even more challenging given the uneven impacts of climate change. Yet elusive global action by all workers may very well be necessary to mitigate and adapt to climate change (Uzzell and Räthzel 2013). While no single project or strategy will mitigate successfully on its own, it is now mainstream to call for concerted efforts across the planet to respond to the most recent science on climate change.

> As the results from the latest and best available science become clearer, the challenge becomes more daunting, but simultaneously the solutions become more apparent. These opportunities need to be grasped across society in mutually reinforcing ways by governments at all levels, by corporations, by civil society and by individuals. (UNFCCC 2013)

The question remains: is it possible to achieve such a scale of action if labour geography's unit of analysis is limited to institutional labour (i.e., unions)? Indeed, the conflation of labour with class is another criticism of labour geography (Das 2012). Labour geographers are more than aware of the challenges posed by the fragmentation of workers by gender, race, class, and geography to building strong movements for social change. When community concerns over the environment and contradictions between blue and green agendas are considered, such unity proves even more elusive (Nugent 2011).

Labour geographers are also aware that the roles of the state and political engagement have been less than fully studied (Rutherford 2013). In terms of trade union retrenchment, public sector unions have arguably become the core of organized labour in many advanced capitalist economies. These unions are, however, increasingly under threat in the current age of austerity as states launch an intensified attack on workers (Jordhus-Lier 2012). The state also regulates nature through laws, policies, practices, and institutions. The role that labour plays in governmental processes, either on its own or in conjunction with allies in the environmental movement, is an area that will continue to challenge unions. Labour geography's conceptualizations of agency, the production of scale, class formation, and the state are therefore further complicated as questions of nature are considered in capital-labour-nature relationships.

CAPITAL-LABOUR-NATURE MATRIX

In an important article that attempts to set out a meta-analysis of possibilities for governance in response to climate change, Wainwright and Mann (2012) develop a matrix that pits capitalist and non-capitalist economic systems against planetary and anti-planetary sovereignty. "Climate Leviathan," the title of the article and perhaps the most nightmarish scenario, sees the emergence of a global capitalist system structured to respond to climate change while maintaining accumulation. Climate Behemoth is another capitalist scenario that posits a world without planetary sovereignty and a range of state responses ranging from populist climate change denial to regional green economic development strategies. Non-capitalist responses are classified as Climate Mao, which suggests the emergence of strong planetary, sovereign-state lead responses; and the unknown Climate X, the anti-sovereign scenario in which possibilities for resistance and innovation reside.

Our approach is not on the ambitious conceptual scale of Wainwright and Mann (2012), but we are influenced by their matrix. Our framework poses labour- and capital-influenced responses to environmental concerns against two degrees of environmental action, one that preserves the absolute sovereignty of capital (e.g., greenwashing) and the other that grants more sovereignty to nature (see Figure 8.1 below). It is important to note that the entire matrix (detailed briefly below) is found within the confines of the capitalist system, although each scenario would have arguably different impacts on accumulation. A second key point is that we believe that instances of true labour environmentalism are rare and often fleeting, although this is perhaps the scenario that presents the strongest possibilities for change.

Management Greenwashing

The term *greenwashing* has its roots in the accommodation sector. It can be traced back to a 1980s essay written by environmentalist Jay Westerveld, who coined the term in response to a hotel's request as to whether he wanted his sheets washed (Hayward 2009). He thought that a hotel promoting a minor environmental practice that actually saved it money

FIGURE 8.1
Capital-Labour-Nature Matrix

Source: Authors' compilation.

was practising a double standard. Management greenwashing persists as firms respond to consumer demands for action on climate change. Yet many actions have marginal, if any, concrete impact on the environment. These initiatives are largely voluntary and unregulated by the state. In some instances, as discussed below, greenwashing is used to discipline labour by reducing the inputs necessary to produce a good or service. In this case, environmental friendliness does little or nothing to disrupt corporate profits and may even increase profitability.

Eco-capitalism

Eco-socialists argue that capitalism is simply incompatible with environmental sustainability (Magdoff and Bellamy-Foster 2010). In the final instance, this is quite possibly the correct assertion. Yet capital is not equally hostile to better environmental practices. While it is a mistake to assume that market-based solutions provide the only answer to environmental crises, capital is not monolithic as economies, sectors, and firms are responding unevenly to climate crisis. Carbon footprints vary a great deal by sector and geography. In some cases, firms can profit through adaptation and mitigation to climate change in highly regulated economies. Some argue that where rationing, monopoly pricing, and green subsidies are present, profitability is possible with the greening of production (e.g., Barbier [2010] on the Green New Deal; Jones 2008). Yet the needs of capital dominate, and the displacement of workers, through processes of restructuring, persists as the underlying contradictions of capital (green or not) persist (Smith 2007).

Union Greenwashing

The idea that labour is a complicit partner in greenwashing may be uncomfortable to some. Yet organized labour often partners with capital in processes that lead to increased sector- or firm-level competitiveness (e.g., lobbying for state subsidies). This type of greenwashing, however, may differ substantially in form from those efforts led by management. For example, shallow coalitions (Tattersall 2010) between labour and environmental groups may form to greenwash and protect jobs. Union greenwashing may also be used, as demonstrated in the cases discussed below, as a means of gaining strategic leverage over employers or governments that fail to present viable environmentally friendly practices. The reforms advocated by unions, upon close examination, are also contradictory and do little for the environment even as they prioritize the narrowly defined needs of workers (e.g., job security, higher wages, training opportunities).

Labour Environmentalism

What would true labour environmentalism within the confines of capitalist accumulation look like? If the needs of workers and nature are to be reconciled in any meaningful way, the science of climate change has to be considered in a more democratic manner and allow workers and unions to exercise power in mitigation and adaptation. Lipsig-Mummé (2013) identifies six points in the chain of production at which workers play a role: inputs into the labour process, the organization of work, new technology, the physical environment in which the work takes place, distribution of goods, and disposal of waste and end of the product's life.

Clearly, there are many points of entry for labour, but to influence these processes, workers must exercise real power. True labour environmentalism would provide a just transition for workers displaced from sectors that produce goods and services that are simply no longer environmentally sustainable (Snell and Fairbrother 2013). At the same time, there would be an emphasis on substituting the amount of non-fossil-fuel-consuming labour for capital in production processes. In order to achieve this, labour must go well beyond greenwashing and look at its role in supporting deep coalitions, global social movements, and governments willing to intervene in reshaping the economy. We are some distance from achieving this scenario.

Presenting our framework in the form of a matrix is somewhat misleading in that it seems to present discrete and absolute ideal types. However, these ideal types are better understood as points on a continuum. They are dynamic relationships as capital, labour, and nature are engaged in constant dialectical tension. Union and management contest each other through different forms of greenwashing at the same time as some labour groups may be in conflict with one another over what constitutes union greenwashing versus a more meaningful labour environmentalism.

Showing how labour is active in producing hospitality and tourism landscapes can certainly benefit tourism and labour geography (Tufts 2013; Zampoukos and Ioannides 2011). With this in mind, we now provide a brief discussion of greenwashing in hospitality as well as green union campaigns in that sector.

GREENWASHING IN THE HOSPITALITY SECTOR

As already noted, for almost two decades, large hotels have initiated environmentally based programs to save money and reduce the inputs of energy and chemicals into the guest experience. Many such programs involve training workers in such responsible environmental behaviour (Gossling 2010). In the 1990s, hotels began asking guests to state whether they were content with their sheets not being changed every night during prolonged stays. Today, this is common practice in most hotels, as is the practice of asking guests to leave unused towels on the towel rack and put used towels on the floor or in the bathtub. These practices reduce not only the energy used to do laundry but also the costs that are associated with it. The growth of environmental consulting firms (e.g., Green Hotels Association, EcoGreenHotel, Green Consultants), which cater to accommodation companies wishing to reduce their costs and environmental footprint, is evidence of how seriously the industry considers green practices.

More recently, hotels have taken this practice and marketized it by sharing the savings directly with consumers. On 18 November 2010, a group of hotel workers and a small delegation of community supporters entered the Sheraton Centre Toronto Hotel to protest the Make a Green Choice program, which gives guests a $5 per night discount if they choose not to have their room serviced.

The hotel company claimed that the savings in energy (and carbon dioxide emissions) and chemicals are beneficial to the environment. The Toronto workers, represented by UNITE HERE Local 75, countered that there was no real reduction in harm to the environment because a room that has not been serviced in days consumes almost the same amount of energy to clean as a room maintained every day. The union argued that the Make a Green Choice program resulted in intensified work for room attendants, who still clean rooms on a quota system, and that it was simply a means of reducing labour costs (UNITE HERE Local 75 2010). Room attendants – mostly migrant women – are the largest group of workers in hotels. The Make a Green Choice program reduces the number of rooms to be serviced (i.e., by decreasing the amount of work for individuals), while increasing the amount of cleaning to do in rooms that have not been serviced every day (i.e., intensifying the labour process). Clearly, the program has significant immediate and future implications for workers, and the real savings to the environment are largely unproven.

The union continues to struggle against the program. Most recently, another demonstration was held at Toronto's Sheraton Centre on 4 December 2014 linked to a global week of action by hospitality workers. One UNITE HERE Local 75 member, Rafunzel Korngut, quoted in the *Toronto Star*, sums up the issue.

> They are using the idea of green choice. But the ethical blindness of this idea is the workers don't have rooms to clean – and that equals, they don't have a job. And then if people don't have a job, they don't have food to bring to the table. (Mojtehedzadeh 2014)

Several green hotel eco-certification programs have been developed (e.g., Green Seal Hotels & Lodging, EarthCheck, EcoLogo, Qualmark, and Green Key Global). Green Key Global's widespread certification program is the Green Key Eco-Rating Program, and it is endorsed and promoted by the Hotel Association of Canada. The Green Key program is voluntary for hotels, motels, and resorts in Canada and the United States and charges an annual registration fee of $350 in Canada ($600 in the US). An online, self-administered audit evaluates several operational areas of a property as well as its energy use, water use, waste management, building infrastructure, land use, environmental management, and indoor air quality. Based on its practices, the facility can obtain a rating of 1 (lowest) to 5 (highest). The Green Key program then certifies the property, sends it a plaque with its rating, and pinpoints it on its interactive online map, along with all other Green Key–certified hotels. The company states on its website that it may conduct an on-site inspection to confirm the rating (Green Key Global 2014).

It is argued that sustainable tourism certification helps to minimize the industry's negative impacts on the environment, cultures, and societies (Rome 2005). A number of certification programs are offered for hotels, ranging from online self-auditing programs to third-party audits (Clarke 2002). Most hotels require an incentive to become certified, such as reduced operating costs, a positive image for their brand or logo, marketing opportunities, or moral justification (Rome 2005). Thus, many certification programs offer the use of their logo, association with their brand, listings in directories and on websites, media attention, and access to green markets (ibid.). As these marketing benefits are realized, an increasing number of hotels are working toward achieving eco-rating certification (Bedlington 2009).

There are, of course, criticisms of such certification programs. Some are little more than paid memberships that provide positive public relations and media attention (WWF-UK 2000). Tourists may find it difficult to discern which certification programs are genuine. Many programs have conditions that are easily accessible and achievable by most businesses. A third-party audit may increase the transparency of the certification, as does the awarding of certification after the environmental commitment is achieved (WWF-UK 2000; Boluk 2011b). Yet third-party certification providers can also be subject to abuse, with no effective auditing process.

Eco-rating certification provides an exclusive competitive advantage to hotels through the use of a recognizable logo (WWF-UK 2000). However, this is problematic as certification can be costly, thereby excluding smaller businesses that may not have the available financial resources. Not surprisingly, these programs cater to larger hotels. Most eco-rating certification programs focus solely on environmental performance and exclude the holistic concept of sustainable tourism. As well, although certification programs focus on environmental performance, this is often related to the structure and use of a building; rarely do eco-rating programs focus on critical contextual environmental issues, such as local biodiversity and habitat loss. Some certification programs, such as Green Globe 21, are more encompassing than most, requiring companies to protect surrounding habitats, ecosystems, and endangered species while educating their guests about these issues (WWF-UK 2000).

Most certification programs offer little guidance about how to integrate social and cultural issues, such as social responsibility and economic equity, into tourism; however, there are a few exceptions (Boluk 2011a, 2011b). Implementing certain environmental actions or performance measures may adversely affect a hotel's employees; thus, it is important for certification programs to include social measurement in the process. This is particularly important if changes that are made to accommodate environmental protection impact the labour process or the quantity of work needed. For these reasons, alternative rating systems are emerging, some created by labour unions in the hotel sector, advocating both labour and environmentally friendly practices.

UNIONS, HOSPITALITY, AND THE ENVIRONMENT

A number of union campaigns have emerged in recent years that integrate labour and environmental issues into the production of accommodation and food services. Such "fair hotel" certification programs sponsored by labour unions are in their infancy, emerging only in the last few years. In Ireland, for example, the Services, Industrial, Professional and Technical Union (SIPTU) launched a Fair Hotels campaign in 2010 "in response to the wide scale denial of workers' rights in the Irish hotel industry as evidenced by official statistics from the State's labour inspectorate" (SIPTU n.d.). In order to receive Fair Hotels certification, a hotel must:

1. Recognise staff's right to collective representation in the workplace.

2. Let staff know that they are free to form a union without intimidation or hindrance.

3. Arrange for staff to meet with Fair Hotels organisers. According to the Irish Hotels Federation 2009 Annual Report, there are now over 900 hotels in Ireland following a period of significant expansion since 2005. To date, 46 hotels are certified as Fair Hotels. (Fair Hotels 2014)

What is unique about the Fair Hotels approach is that while the aim of the campaign is to improve the quality of employment and workers' rights in the sector, the strategy is largely based on a positive boycott. In many ways, the program aims at increasing the competitive advantage of hotels that are less hostile to workers. In the initial stages, the focus was simply placed on convincing workers and unions to patronize Fair Hotels, but this strategy has been extended to include community groups (Buckley 2012). Fair Hotels seeks to direct business (individuals, conference planners) to certified hotels through its website. Given the relative infancy of the program, little research has been done on its impact.

Karla Boluk (2013) has interviewed Fair Hotels management and found that while the corporate social responsibility designation has notably increased business and value for employers and workers, it has not necessarily changed overall employee satisfaction. What is also evident in the campaign is its pure focus on industrial relations and collective bargaining rights but the absence of any other criteria, including environmental practices. There are, however, examples of union campaigns that have integrated environmental criteria into their pro-worker campaigns.

Australia and United Voice's First Star Program

The Australian hotel industry also has relatively low union density. In 2012, less than 5 percent of accommodation and food services were unionized in Australia, a decrease from almost 20 percent in the mid-1990s and 9 percent in the mid-2000s (Peetz 2005; ABS 2013). The Liquor, Hospitality and Miscellaneous Workers Union (LHMU, renamed United Voice in 2011), has a limited presence in the sector. LHMU launched the First Star program in 2009 as a means of raising standards and reducing labour turnover in the sector.[2] It differed significantly from Ireland's Fair Hotels campaign because the rating system incorporated environmental as well as fair labour practice criteria. The First Star vision for an environmentally sustainable hotel included an honest commitment to no greenwashing, engaged workers and guests (in environmentally friendly practices), energy and water efficiency, respect for the local environment, and adherence to government standards and goals for reducing greenhouse gas emissions. It is also noteworthy that the union is part of Climate Action Network Australia.

In February 2011, interviews were held with union officials responsible for the First Star campaign. The interviews revealed that the program had been largely developed from the top down; the union had outsourced its development to a media firm, initially as a labour-management partnership to encourage hotels to improve retention through better employee relations and to attract guests. Notably, First Star was slow to award its first designation, but after two years, it went to IHG Hotels, with much fanfare, which included government and industry representatives. Up to that point, the union had deemed all Australian hotel chains to be unworthy of a First Star rating. Louise Tarrant, national secretary of United Voice, when awarding the designation, stated,

> United Voice is extremely proud to be working with IHG to ensure quality hotels across Australia....
>
> As anyone who has stayed in an Intercontinental, Crowne Plaza or Holiday Inn can tell you, these are truly quality hotels worth staying at. But, no one knows hotels better than the workers who work inside of hotels. Hotel workers have inside knowledge of what's going on behind the scenes – of what hotel operators are actually doing.
>
> And it is workers at IHG hotels who tell us that IHG is doing the right thing – by the environment and by their staff. That's why they are being awarded a First Star rating. (*ETB Travel News* 2011)

Despite the initial fanfare, it appears that the First Star campaign has run its course, and the website is defunct. Indeed, at the award ceremony, it was stated that the LHMU was kicking off the Better Jobs, Better Hotels campaign, which was actually the name of a campaign it had launched in 2008.

In the interviews, union staff conceded that the union leadership had not been happy with the launch of First Star and had been reconceptualizing the campaign even before it awarded its first designation. At the time, United Voice had shifted its focus toward its more recent Hotels with Heart campaign, which was aimed at raising standards in the sector through more traditional means (e.g., bargaining) and shaming employers with poor employment relations with reports such as *Heartbreak Hotels* (VIRWC 2010). The program is very much inspired by UNITE HERE's Hotel Workers Rising campaign in North America, and it represents a shift away from a consumerist strategy toward applying direct pressure on employers and the government. While the First Star program included criteria beyond working conditions, the success of a sustained program has proven elusive.

North America and UNITE HERE's Real Food Real Jobs Campaign

Union density in hotels in North America is difficult to convey meaningfully. National union density in the hotel sector is lower than the all-industrial average, but union membership is mostly concentrated in large, full-service hotels in metropolitan centres. For example, in Canada, only 16.7 percent of hotel workers are unionized (just over half the average for all workers), yet in Toronto, the majority of workers in the large downtown hotels are unionized (Tufts 2007). Similarly, hotel workers in Las Vegas are also highly unionized (Gray 2004). In this respect, the case is slightly different from Ireland and Australia.

UNITE HERE is a major hospitality union in North America dating back to the 1890s; it now represents over 270,000 workers, mostly in hotels, but with 90,000 members in food services. The union has historically used consumer boycotts against hotels and, over the last decade, has adopted information and communications technology to develop its Union Hotel Guide, first as a web-based search engine for finding union-friendly hotels and more recently as a smart phone application. In the United States, UNITE HERE has also established the Informed Meetings Exchange (INMEX), a web-based utility that assists meeting planners with socially responsible event planning. Founded in 2006, the non-profit organization alerts meeting planners to upcoming potential work stoppages in hotels; promotes *force majeure* clauses in contracts, which allow events to be moved during a strike; and even provides assistance with logistics in order to attract business to union hotels. While the INMEX platform was publicized at its inception, it has been largely stagnant in recent years. In interviews, a UNITE HERE official offered several explanations for this. Organizational capacities for the entire union shifted toward the Hotel Workers Rising campaign, and there was a vicious inter-union conflict. And following the election of Barack Obama in 2008, key union activists who had developed INMEX left the union to work in the public service. (However, it was reported that INMEX would receive more resources from the union in the near future.)[3]

The Union Hotel Guide and INMEX are aimed solely at the labour-friendliness of hotels. Like the Fair Hotels campaign in Ireland, there is no consideration or even mention of green criteria. There are, however, obvious opportunities for UNITE HERE to intervene in the green certification process through INMEX and its Union Hotel Guide. Specifically, the union itself could rate its employers and issue a rating similar to the well-known Diamond and Star ratings used by travel providers. The precedents for this type of action, discussed above, clearly demonstrate possibilities, although not guaranteed success.

While the union has not integrated green strategies into its hotel-consumer-patronage plans, environmental issues are evident in its more recent efforts to raise standards in the institutional food services sector. In 2012, the union launched its Real Food Real Jobs campaign. This is a comprehensive strategy based on three principles (see Figure 8.2 below). The first, Real Food, reflects the green elements of the campaign, advocating for healthy meals and the ethical and local sourcing of foods. The Real Jobs component (shown in bold font in the campaign logo) represents the call for higher standards for workers and democratic rights in the workplace. The final principle, Transparency, is directed at institutional employers, demanding full disclosure of food sourcing and employment policies. The logo itself is a graphical summary of these three principles: workers' rights (clenched fist), healthy food (fork), and local sourcing (green leaf).

FIGURE 8.2
Real Food. Real Jobs.

What do we stand for, and what do we want?

1. **Real Food:** We support a food system that emphasizes fresh cooked meals rather than processed items, prioritizes the local and ethical sourcing of ingredients, and utilizes production methods that are humane and respect our environment.

2. **Real Jobs:** Food workers should be paid a living wage (with health and retirement benefits), including enough to afford real food for their families. Workers should be free to publicly disclose food safety or quality issues, and to form a union through a legal and democratic process of their own choosing without threats and intimidation.

3. **Transparency:** Transparency is fundamental in changing our food system. Universities – and all food service institutions – should fully disclose the source of their food purchases, and the wages and benefits paid to food workers.

Source: Real Food. Real Jobs. (http://www.realfoodrealjobs.org/principles).

The Real Food Real Jobs campaigns are aimed at public schools, universities, and airports. While the focus on food services in the education sector is an important campaign, given the tourism-related focus of this chapter, we focus here on the campaign to green airport concessions. Real Food Real Jobs is reportedly based in North America, but spatially it is concentrated in selected cities. The major concentration of airport activity is in Chicago, built around UNITE HERE Local 1's efforts to secure union jobs in the midst of a $3 billion upgrade of O'Hare International Airport (ORD). The campaign at ORD follows a long history of Local 1 organizing concession workers there. The timing of the campaign was, however, also related to the massive investment at ORD and the opportunity to organize members at the new concessions. The union's stronger presence at Los Angeles International Airport allowed it to protect its work more easily during the $4 billion upgrade there.

The strategy of Local 1 and the Real Food Real Jobs campaigns is to build community support in their fight to organize the workers of new concessions developed throughout ORD's upgrade. The campaign, launched in early 2012, began with a survey that asked consumers what type of food services they demanded at the airport. The union highlighted the demands of travellers for local products in airports as a best practice, inspired in part by the 100-Mile Diet and other Eat Local campaigns. Another consumer-based strategy was a positive boycott that attempted to direct consumers to unionized, high-quality concessions. One campaign attempted to intervene in a *USA Today* Readers' Choice "10 best airport restaurants" poll by encouraging customers to vote for the unionized nominees.

UNITE HERE also involves the Real Food Real Jobs campaigns with its members and the community. Workers feature as advocates for locally grown food and unionized jobs, and member activists are sent to events organized by the local food security community to talk about how airport sourcing and employment practices can fundamentally change the food system. Community engagement is a significant part of the campaigns.

In the spring of 2013, Local 1 released its key report, which advocated for sustainable food service practices at ORD and listed its key demands. In "Putting Sustainability on the Table," the union makes its case for standards at the airport, drawing on findings from a survey of 200 airport concession workers. Specifically, its demands were:

1. ORD-er fresh, healthy, and local food (10 percent from the US Midwest).

2. Limit food waste through better management and by donating surplus food.

3. Make real food work by creating living-wage unionized jobs and a peaceful transition as new concessions are awarded. (UNITE HERE Local 1 n.d.)

The report was not simply promoted through press releases; the union also organized several events and action around its release. It was recognized that the powerful Chicago Department of Aviation (CDA) would need to face public pressure before it responded to the demands of the report. In April 2013, the CDA announced its Green Concessions Policy for ORD (which the union took as a first victory in its ongoing struggle), even adopting it on 21 April in honour of Earth Day.

In 2011, the CDA had established its own certification program for tenants, the Green Airplane Rating System (CDA 2013). It required concessions to use more sustainable packaging and fewer disposable utensils and to even donate surplus food; however, there was no mention of labour standards or guarantees of union recognition. Questions remain about the future overall environmental impact of UNITE HERE's demands (e.g., uptake, perceived versus real benefits of localism) and, more important, the long-term compatibility of green and blue demands in the sector.

DISCUSSION: RAISING STANDARDS OR UNION GREENWASHING?

The cases discussed above are admittedly experimental and uneven in how they integrate environmental criteria and material gains for workers. It is possible, however, to reflect on their limits and potential for hotel workers, the sector, and the environment. There is little evidence that these campaigns reflect any significant form of labour environmentalism. In both the First Star program and the Real Food Real Jobs campaigns, environmental criteria are present, but they do not in any way challenge a firm's ability to profit. In fact, it can be argued that the unions are assisting a firm's marketing efforts by providing environmental designations to attract people to hotels and airport concessions. In her assessment of Fair Trade in Tourism South Africa, Boluk (2011a, 2011b) has argued that ethical consumption of hotels and other tourism-related services is quite consistent with conspicuous consumption under capitalism as consumers valorize the moral self through their consumption choices. Hospitality workers' unions are simply attempting to turn this to their advantage by integrating appeals to consumers to consider working conditions and environmental practices in their consumption decisions. In part, this can be a promising strategy, but there are also limits.

These are all boycott strategies, even if they are positive boycotts in some cases. The long-term effectiveness of union-led consumer boycotts has been questioned for some time (Meyer and Pines 2005), although a full discussion is beyond the scope of this chapter. The point here is that a positive boycott may build some public support for unions, which are clearly in a defensive position with both employers and consumers. At the same time, unions play an adversarial role with employers and in time of conflict will retreat to a negative boycott strategy – at the heart of which is the strike or lockout. It is difficult to shift relations with any legitimacy from "please patronize" to "please boycott" with every bargaining cycle.

There are, however, clear benefits to having unions involved in green certification processes. While not completely independent, a union-endorsed certification has arguably more legitimacy than a rating from a fee-for-service provider. When a union or a union-created front organization (i.e., an organization created, controlled, and supported by unions) certifies a "green" business, it may be viewed as a legitimate party, neither government- nor purely industry-based. This legitimacy would allow hotels (at least those willing to participate) to avoid accusations of greenwashing their services through self-reporting systems, something customers view with increasing skepticism.

Even though a union may be considered a legitimate party for awarding certification, the currently limited scope and scale of such initiatives present some challenges. In some cases, certification is restricted to issues of collective bargaining rights; in others, environmental criteria are included, but it is less clear who has developed such criteria or how. There is also an important question about the scale of the campaigns. Are small operators able to meet the same social and environmental criteria as large firms? In the geographical sense, at what scale should a rating system be developed: locally, regionally, nationally, or internationally? Would a rating system developed by Anglo-American unions be fairly applied to hotels in the Global South?

Here the paradox of climate change becomes apparent. Specifically, while there is broad consensus that climate change is a global phenomenon that requires global action, climate change is also a geographically uneven process that inspires different degrees of action (Swyngedouw 2006). For example, climate change means much more to tourism workers in small Pacific island states facing rising sea levels and beach erosion than hotel workers in continental North America (Milne 2013). Such varied effects will challenge international solidarity among hotel workers and their unions. It is also too early to tell how these campaigns will flow through different jurisdictions. For example, the Real Food Real Jobs campaign, largely developed in the context of Chicago, may fail to live up to the North American aspirations of UNITE HERE. It will take some time for these campaigns to scale up to truly global, labour-driven initiatives or even expand to airports where UNITE HERE does not represent concession workers. But it is important to note that broad farm-to-table movements have been expanding geographically for some time, and they have popular appeal (Pollan 2006).

As we have seen, unions can incorporate social and industrial relations criteria into green rating systems (e.g., community involvement of a firm, neutrality in organizing practices) and vice versa. Integration is important as divorcing the social (in this case, work issues) from the environmental can lead to a very eco-centric environmentalism. For example, recycling in hotels is good, but how sustainable is it if work only intensifies for room attendants forced to sort waste? It is here, in looking at how changes in work may very well be the centre of effective climate change adaptation and mitigation, that Rutherford's (2010) call for a renewed emphasis on the labour process in labour geography may also be warranted (see also Lipsig-Mummé 2013). At the same time, unions' consideration of environmental issues strengthens their position with consumers and communities as it draws attention to conditions beyond the narrow issues of wages and working conditions.

It is obvious that unions are seeking to develop ratings and certifications that can be used to play employers against one another (union versus non-union) in the competitive process. Yet labour as capital depends on capital circulation, and both hotel workers and hotel owners (especially small proprietors) are harmed when tourists choose other hotels or even entire locales. Thus, any certification system has to reward a hotel quickly with a higher rating for compliance; otherwise, the entire firm is jeopardized. In some cases, unions may be just as tempted to rate a hotel more favourably as the hotel company itself. Despite their differences, the campaigns we have discussed are embedded in what Räthzel and Uzzell (2011) identify as a mutual interests discourse in which management, workers, and the environment can benefit through co-operation and solidarity. In this approach, unions may in fact distance themselves from deeper partnerships with communities and broad social movements that seek more significant economic and environmental change.

In order to overcome these problems and gain legitimacy for their rating systems in the short term, unions have to build trust with environmental organizations. Indeed, the promise here is that such systems will necessitate broad coalition building, as is the case with the campaigns discussed above. There will, however, inevitably be tensions over how to weigh green against blue criteria in rating systems. Such contradictions remain persistent in any blue-green alliance. Using the Make a Green Choice program as an example (see "Greenwashing in the Hospitality Sector" above), reducing the demand for clean towels and sheets does reduce the demand for energy and labour, but if jobs are lost in the process, will unions judge environmental protections to be less important than fair criteria?

The purpose of unions within capitalism is to protect and enhance jobs, and campaigns that advocate for workers to increase their skills (e.g., waste management in food services, cleaning with fewer chemicals) are warranted. In the auditing and certification process, for example, workers are presented with learning and training opportunities. If unions are going to certify 'green' hotels, union activists could do these audits themselves after receiving the proper auditing training and identifying opportunities to reduce a hotel's carbon footprint. In many instances, workers already do this by participating in audits of accessibility and health and safety. Like health and safety, the environment may become an arena for engagement and negotiation, one in which labour and management can build structures for less adversarial workplace relationships. Conversations with UNITE HERE officials about implementing a future worker auditing program identified the real possibility of setting up health and safety committees.

CONCLUSION

There are, however, hard limits to campaigns such as those briefly touched on here. There is always the danger of co-opting union certification toward a greenwashing outcome, but more relevant is whether these campaigns allow unions to face the difficult questions about the limits of the relationship between expansionary capitalism and nature. At the moment, green fair certification programs are largely techno-centric responses to environmental challenges. Such certification practices are less likely to produce a true labour-environmentalist policy response to interventionist strategies such as rationing air travel, which reduces overall tourism activity and would require a truly just transition for marginalized workers in tourism-related industries (Monbiot 2006).

Such fair trade certification has been criticized as reformist, but this does not mean that such initiatives will have no tangible environmental benefit or impact on climate change mitigation and adaptation. If capitalism and nature are in a dialectical relationship, then we must regard both capital and labour as implicated (Castree 2001, 2002; Harvey 1996; Swyngedouw 2006); the environment does affect how we interpret labour's agency in processes such as greenwashing and fair trade. It is also clear that labour scales campaigns and builds broad class coalitions in distinct ways in response to environmental concerns. While the state has not been a focus of this chapter, it is clear that the absence of big government in regulating certification has itself created a vacuum of self-regulation, which hospitality unions are beginning to enter.

Tourism and hospitality landscapes will continue to develop unevenly. Exploring labour's role in producing tourism geographies that are also affected by nature is an approach worthy of further development. While no cases of true labour environmentalism are identified in the campaign cases presented here, studying the tensions among emerging scenarios is one way of understanding how tourism workers and industries are responding to the changing environment.

NOTES

1. For a definition of this term, see the "Management Greenwashing" section.
2. Interviews with United Voice campaign staff, 15 February 2011, Sydney, Australia.
3. A UNITE HERE official who helped develop INMEX introduced it to an academic conference on hospitality work held in Toronto, 13 October 2006. The same official eventually left to work in the Obama administration (interview with UNITE HERE official, 25 March 2011, Toronto).

REFERENCES

ABS (Australian Bureau of Statistics). 2013. *Employee Earnings, Benefits and Trade Union Membership, Australia*. Cat. no. 6310.0. Canberra: ABS. http://www.abs.gov.au/ausstats/abs@.nsf/mf/6310.0.

Barbier, E. 2010. *A Global Green New Deal: Rethinking the Economic Recovery*. Cambridge: UNEP / Cambridge University Press.

Bedlington, E. 2009. "Where Is the Value in Greening Your Brand? An Analysis of the Canadian Hotel Industry." Master's thesis, Simon Fraser University, Burnaby, BC.

Boluk, K. 2011a. "Fair Trade Tourism South Africa: Consumer Virtue or Moral Selving?" *Journal of Ecotourism* 10 (3):235–49.

—. 2011b. "In Consideration of a New Approach to Tourism: A Critical Review of Fair Trade Tourism." *Journal of Tourism and Peace Research* 2 (1):27–37.

—. 2013. "Using CSR as a Tool for Development: An Investigation of the Fair Hotels Scheme in Ireland." *Journal of Quality Assurance in Hospitality & Tourism* 14 (1):49–65.

Buckley, L. 2012. "Fair Hotels Campaign Ireland." Presentation to Union Experience in Hotel Labour Markets, Oslo, 14 June.

Burkett, P. 1999. *Marx and Nature: A Red and Green Perspective*. New York: St. Martin's Press.

—. 2006. *Marxism and Ecological Economics: Toward a Red and Green Political Economy*. Boston: Brill.

Carey, J., and S. Tufts. 2014. "'Greening Work' in Lean Times: The Amalgamated Transit Union and Public Transit." *Alternate Routes* 25:207–33.

Castree, N. 2001. "Socializing Nature." In *Social Nature: Theory, Practice and Politics*, ed. N. Castree and N. Braun, 1–21. Oxford: Blackwell.

—. 2002. "False Antithesis? Marxism, Nature and Actor Network Theory." *Antipode* 34 (1):111–46.

—. 2007. "Labour Geography: A Work in Progress." *International Journal of Urban and Regional Research* 31 (4):853–62.

Castree, N., N. Coe, K. Ward, and M. Samers. 2004. *Spaces of Work: Global Capitalism and Geographies of Labour*. Oxford: Blackwell.

CDA (Chicago Department of Aviation). 2013. "CDA Unveils Green Concessions Policy for O'Hare and Midway International Airports." Press release, 21 April. http://www.flychicago.com/business/EN/media/news/stories/pages/NewsDetail.aspx?ItemID=844.

Clarke, J. 2002. "A Synthesis of Activity Towards the Implementation of Sustainable Tourism: Ecotourism in a Different Context." *International Journal of Sustainable Development* 5 (3):232–50.

Coe, N., and D. Jordhus-Lier. 2011. "Constrained Agency? Re-evaluating the Geographies of Labour." *Progress in Human Geography* 35 (2):211–33.

Crutzen, P.J., and E.F. Stoermer. 2000. "The 'Anthropocene.'" *Global Change Newsletter* 41:17–18.

Das, R. 2012. "From Labor Geography to Class Geography: Reasserting the Marxist Theory of Class." *Human Geography* 5 (1):19–35.

ETB Travel News. 2011. "InterContinental Hotel Group and United Voice Launch the First Star." 13 October. http://australia.etbnews.com/117821/intercontinental-hotel-group-and-united-voice-launch-the-first-star/.

Fair Hotels. 2014. "Welcome." www.fairhotels.ie/.

Gossling, S. 2010. *Carbon Management in Tourism: Mitigating the Impacts of Climate Change*. Abingdon, UK: Routledge.

Gray, M. 2004. "The Social Construction of the Service Sector: Institutional Structures and Labour Market Outcomes." *Geoforum* 35 (1):24–34.

Green Key Global. 2014. "Green Key Eco-Rating System." http://greenkeyglobal.com/programs/eco-rating-program/.

Harvey, D. 1996. *Justice, Nature and the Geography of Difference.* Oxford: Blackwell.

Hayward, P. 2009. "The Real Deal? Hotels Grapple with Green Washing." *Lodging Magazine*, 1 February. http://web.archive.org/web/20090205171221/http://www.lodgingmagazine.com/ME2/dirmod.asp?sid=&nm=&type=Publishing&mod=Publications%3A%3AArticle&mid=8F3A7027421841978F18BE895F87F791&tier=4&id=FD212DB2AA944808BF5CE6519B2BCC06.

Herod, A. 1997. "From a Geography of Labor to a Labor Geography." *Antipode* 29 (1):1–31.

—, ed. 1998. *Organizing the Landscape: Geographical Perspectives on Labor Unionism.* Minneapolis: Minnesota University Press.

—, ed. 2001. *Labor Geographies.* New York: Guilford Press.

—. 2010. "Labour Geography: Where Have We Been? Where Should We Go?" In *Missing Links in Labour Geographies*, ed. A. Cecilie Bergene, S.B. Endresen, and H. Merete, 15–28. Surrey, UK: Ashgate.

—. 2011. *Scale.* London: Routledge.

Jones, V. With A. Conrad. 2008. *The Green Collar Economy: How One Solution Can Fix Our Two Biggest Problems.* New York: HarperCollins.

Jordhus-Lier, D. 2012. "Public Sector Labour Geographies and the Contradictions of State Employment." *Geography Compass* 6/7:423–38.

Lier, D. 2007. "Places of Work, Scales of Organising: A Review of Labour Geography." *Geography Compass* 1 (4):814–33.

Lipsig-Mummé, C. 2013. "Climate, Work and Labour: The International Context." In *Climate@Work*, ed. C. Lipsig-Mummé, 21–40. Halifax: Fernwood Publishing.

Magdoff, F., and J. Bellamy-Foster. 2010. "What Every Environmentalist Needs to Know about Capitalism." *Monthly Review.* http://monthlyreview.org/2010/03/01/what-every-environmentalist-needs-to-know-about-capitalism.

Meyer, D., and G. Pines. 2005. "Stopping the Exploitation of Workers: An Analysis of the Effective Application of Consumer or Socio-political Pressure." *Journal of Business Ethics* 59 (1/2):155–62.

Milne, S. 2013. "Climate Change, Tourism and Work: South Pacific Micro-states in Perspective." Paper presented at the Work in a Warming World (W3) International Conference, "Labour, Climate Change, and Social Struggle," York University, Toronto, 30 November.

Mojtehedzadeh, S., 2014. "Hotel Workers Protest Sheraton Centre's 'Green' Program." *Toronto Star*, 4 December, GT3.

Monbiot, G., 2006. *Heat: How to Stop the Planet from Burning.* Toronto: Doubleday Canada.

Nugent, J. 2011. "Changing the Climate: Ecoliberalism, Green New Dealism, and the Struggle over Green Jobs in Canada." *Labour Studies Journal* 36 (1):58–82.

Peetz, D. 2005. "Trend Analysis of Union Membership." *Australian Journal of Labour Economics* 8 (1):1–24.

Pollan, M. 2006. *The Omnivore's Dilemma: A Natural History of Four Meals.* New York: Penguin Press.

Prudham, S. 2005. *Knock on Wood: Nature as Commodity in Douglas Fir Country.* New York: Routledge.

Räthzel, N., and D. Uzzell. 2011. "Trade Unions and Climate Change: The Jobs *versus* Environment Dilemma." *Global Environmental Change* 21 (4):1215–23.

Rome, A. 2005. *Current Range of Incentives Offered to Businesses by "Green" Certification Programs and Quality-Ratings Systems*. Washington, DC: Center on Ecotourism and Sustainable Development, International Ecotourism Society.

Rutherford, T. 2010. "De / Re-centring Work and Class? A Review and Critique of Labour Geography." *Geography Compass* 4 (7):768–77.

—. 2013. "Scaling Up by Law? Canadian Labour Law, the Nation-State and the Case of the British Columbia Health Employees Union." *Transactions (Institute of British Geographers)* 38 (1):25–35.

SIPTU (Services, Industrial, Professional and Technical Union). n.d. "What Is a Fair Hotel?" http://www.siptu.ie/campaigns/siptuorganisingcampaigns/fairhotels/.

Smith, N. 1984. *Uneven Development: Nature, Capital and the Production of Space*. Oxford: Blackwell.

—. 2007. "Nature as Accumulation Strategy." *Socialist Register* 43:16–36.

Snell, D., and P. Fairbrother. 2013. "Just Transition and Labour Environmentalism in Australia." In *Trade Unions in the Green Economy: Working for the Environment*, ed. N. Räthzel and D. Uzzell, 146–61. New York: Routledge.

Swyngedouw, E. 2006. "Circulations and Metabolisms: (Hybrid) Natures and (Cyborg) Cities." *Science as Culture* 15 (2):105–21.

Tattersall, A. 2010. *Power in Coalition: Strategies for Strong Unions and Social Change*. Ithaca, NY: Cornell University Press.

Tufts, S. 2007. "Emerging Labour Strategies in Toronto's Hotel Sector: Toward a Spatial Circuit of Union Renewal." *Environment and Planning A* 39 (10):2383–404.

—. 2013. "Tourism, Climate Change and the Missing Worker: Uneven Impacts, Institutions and Response." In *Climate@Work*, ed. C. Lipsig-Mummé, 141–60. Halifax: Fernwood Publishing.

Tufts, S., and L. Savage. 2009. "Labouring Geography: Negotiating Scales, Strategies and Future Directions." *Geoforum* 40 (6):945–48.

UNFCCC (UN Framework Convention on Climate Change). 2013. "UNFCCC Executive Secretary Christiana Figueres: Latest IPCC Findings a Clarion Call for Global Community to Accelerate Efforts to Combat Climate Change and Steer Humanity Out of Danger Zone." Press release, 27 September. http://unfccc.int/files/press/press_releases_advisories/application/pdf/pr20132709_ipcc.pdf.

UNITE HERE Local 1. n.d. "Putting Sustainability on the Table: Airport Workers' Vision for $3 billion of Food and Drink at O'Hare." Chicago: UNITE HERE Local 1. http://www.realfoodrealjobs.org/wp-content/uploads/ORD-Report.pdf.

UNITE HERE Local 75. 2010. "Toronto Women Leaders Offer Help to Housekeepers in Protest of 'Fake Green Programs' at Sheraton Centre." Press release, 18 November. http://www.uniteherelocal75.org/2010/11/toronto-women-leaders-offer-help-to-housekeepers-in-protest-of-fake-green-programs-at-sheraton-centre/.

Uzzell, D., and N. Räthzel. 2013. "Local Place and Global Space: Solidarity across Borders and the Question of the Environment." In *Trade Unions in the Green Economy: Working for the Environment*, ed. N. Räthzel and D. Uzzell, 241–56. New York: Routledge.

VIRWC (Victorian Immigrant and Refugee Women's Coalition). 2010. *Heartbreak Hotels: The Crisis inside Melbourne's Luxury Hotels*. Melbourne: VIRWC.

Wainwright, J., and G. Mann. 2012. "Climate Leviathan." *Antipode* 45 (1):1–22.
WWF-UK (World Wildlife Fund – United Kingdom). 2000. *Tourism Certification: An Analysis of Green Globe 21 and Other Tourism Certification Programmes*. Washington, DC: WWF.
Zampoukos, K., and D. Ioannides. 2011. "The Tourism Labour Conundrum: Agenda for New Research in the Geography of Hospitality Workers." *Hospitality and Society* 1 (1):25–45.

CHAPTER 9

CITIES, CLIMATE CHANGE, AND THE GREEN ECONOMY[1]

STEPHEN McBRIDE, JOHN SHIELDS, AND STEPHANIE TOMBARI

INTRODUCTION

Climate change is a pressing policy issue, one that has immediate and long-standing impact on the social, physical, and economic environment. Rising rates of urbanization, rapid population growth in some regions, and deteriorating environmental sustainability have challenged governments to respond to these environmental concerns (Schreurs 2008; Corfee-Morlot et al. 2009; Bai 2007). The policy area represents an intricate intersection of multiple social, economic, political, and cultural issues (Gore 2010; Demerse 2011). The impact of the dominant neoliberal policy environment has had a profound effect, at many levels, on how we look at global warming and how we address potential policy responses to green jobs and economic growth. The difficulty of effective action is compounded by the effects of the financial and economic crisis, which, in many countries, has led to renewed bouts of austerity, with obvious implications for governments' ability to invest in new programs.

Canada has a poor record internationally of meeting these challenges. In late 2013, two international reports gave Canada exceptionally poor rankings for its environmental policy. One ranked Canada 55th out of 58 countries in national efforts to reduce greenhouse gas (GHG) emissions (Burck, Marten, and Bals 2013). The other, from the Washington-based Center for Global Development, ranked Canada last among the 27 countries it had rated on the environment (*Maclean's* 2013). Canada

Work in a Warming World, edited by Carla Lipsig-Mummé and Stephen McBride. Kingston: School of Policy Studies, Queen's University.

has been identified as the leader of the global warming villains – a small group of countries opposing environmental reforms and even thwarting international negotiations on the issue (Aulakh 2013). The withdrawal of Canada from the Kyoto Protocol following the United Nations Climate Change Conference in 2011 was a clear statement by the federal government that climate was not a priority. Despite the fact that Canada is one of the world's worst GHG emitters, the Harper government's call for a "Made in Canada" climate change strategy signalled a hands-off approach to federal responsibility.

At the provincial level, there has been some action to limit GHG emissions, sometimes as go-it-alone endeavours, sometimes in co-operation with other provinces and cross-border US states (NRTEE 2012). Some of the more promising efforts, however, have run into institutional obstacles, as was the case with the decision of the World Trade Organization (WTO) to overturn the local content provisions of Ontario's Feed-in-Tariff (FIT) instrument under its *Green Energy and Green Economy Act* (GEA) (Sinclair and Trew, Ch. 1 this volume; McBride and Shields 2013). As well, the area is one where federal-provincial collaboration would be effective, but, as the National Round Table on the Environment and the Economy drily noted, "It is not apparent" (NRTEE 2012, 41).

Unsurprisingly in these circumstances, considerable attention has fallen on Canada's municipalities as possible jurisdictions of last resort in addressing climate issues and simultaneously seeking to enhance a green economy that would also deliver jobs. Yet cities are often viewed as relatively powerless and the least influential level of government; this is because of, in part, a status of subordinate constitutional authority and issues of limited fiscal and other resource capacities. Notwithstanding this view, there have been efforts to argue that cities can play an important role (Katz and Bradley 2013; Toly 2008; Bradford 2002), and understanding the degree to which this is perceived to be true is one objective of this chapter. We first review the arguments of those positing a dynamic role for cities in this area; then we turn to the views of those who are more cautious about cities' potential.

Several subsidiary arguments are advanced in the optimistic literature. Cities are the centres of economic activity and the largest producers of pollution. Cities are also the order of government closest to the citizenry, and cities experience the most immediate and long-standing effects of climate change. Others argue that by recognizing the co-benefits of a green economy, cities have the opportunity to lead the future of climate-change-policy development (Miller 2012). In these ways, cities are strategically located; and hence, cities are compelled to take the lead in constructing a green economy, increasing the number of green jobs, and thus enhancing green growth (ibid.).

For its part, the Federation of Canadian Municipalities (FCM) argues that cities have access to a number of levers, which allow them to tackle climate change in an effective and efficient manner using municipal policy toolkits (Thompson and Joseph 2011). For instance, energy-efficient retrofits, energy conservation, and sustainable water management are policy areas that cities have a direct hand in controlling and managing. Using various other levers, including user fees, property tax exemptions, and zoning policies, cities can support green economic development (Thompson and Joseph 2011; Robinson and Gore 2005). Moreover, by coordinating and managing issues such as mass transit and other transportation, zoning, land use, and planning, it is claimed that cities have the ability to refocus the issue of climate change (Gore 2010; Thompson and Joseph 2011; Kousky and Schneider 2003). Similarly, cities may be well placed to foster public participation and buy-in of a green economic agenda. However, in order for cities to actively pursue green economic development policies and programs, it is widely acknowledged that the "jobs versus the environment" narrative must be challenged (Forstater 2006, 58).

In contrast to these optimistic views of cities' potential, it can be argued that the ability of municipalities to derive, implement, and monitor climate change policies and green economic development is often negated or undermined (Gore 2010). Municipalities in Canada are creatures of the provinces, with limited formal and legal powers; structural constraints and budgetary limitations are significant barriers. Robinson and Gore categorized cities as action and non-action municipalities; however, budgetary restrictions were a barrier to action on climate change proposals even for 47 percent of action municipalities that were leaders in green initiatives (2005). The economic crisis of 2008 and subsequent recession, and the austerity agendas absorbed into municipal budgets, have placed additional restraints on local governments' ability to invest in greening the local economy.

With growing uncertainty about the role that local governments can play in mitigating climate change and promoting green growth, cities often find themselves at a standstill, unable to act because of capacity and informational barriers (Robinson and Gore 2005; Schreurs 2008). Constraints imposed by federal and provincial jurisdictional agreements and divisions of power cut municipalities out of much of the decision-making. However, while cities have traditionally had limited power, neoliberalism may be increasing it, at least by default.

> Devolution and decentralization, celebrated in the canons of new public management, have shifted attention from the national to subnational governments. Such downscaling is reinforced by the growing interests in city-regions as the scale best able to capitalize on newly important economies of agglomeration. (Mahon, Andrew, and Johnson 2007, 43)

Put differently, cities have become sites where neoliberal initiatives intensify the demands and difficulties faced by local governments (see, e.g., Brenner and Theodore 2003), yet may sometimes stimulate inventive contestation (Leitner, Peck, and Sheppard 2007).

With the deflating of the comprehensive Dion-Watson plan in 2006, which aimed to take a "whole of government approach" to climate change in Canada and around the world (Calamai 2007, 40), the abject failure of Stéphane Dion's proposed Green Shift policy in the 2008 federal election, and the withdrawal from Kyoto, the federal government has unabashedly rejected its role in reducing climate change. Moreover, provincial governments vary in their degree of support for GHG reduction incentives and green technology investments (David Suzuki Foundation 2012). As a result, federal and provincial governments may have unofficially increased the flexibility of cities to set policies simply because these other levels of government prefer not to deal with climate change issues themselves.

Still, the lack of a national framework to support the role of cities in greening the economy inhibits their ability to create long-term, systemic change. While cities may face less ideological gridlock than the federal and provincial governments, the absence of collaboration with other levels of government, the private sector, and/or the non-profit sector means that they often do not have the social, economic, and political resources to deliver large-scale and long-term green economic policies (Thompson and Joseph 2011). With this limited authority, cities find it difficult to use policy tools to implement revenue-raising mechanisms, such as user fees and property tax reform, to mitigate climate change and strengthen the local economy. Further disabling is that cities face great public resistance to increased taxes and other municipal charges and fees (Thompson and Joseph 2011; Toly 2008).

In this vein, civic participation can cut both ways. As well as enhancing progress in greening the local economy, it also has the power to impede it by erecting potential barriers. Resistance from the citizenry can inhibit the effectiveness of climate change initiatives and the overall sustainability of green policies. Portney argues that the "three deadly sins" – tragedy of the commons, NIMBYism, and an expanding ecological footprint – can also lie at the heart of unsustainability (2005, 585–86). Similarly, Bulkeley notes that a lack of "theoretical enthusiasm" can inhibit green economic policies from gaining support from the citizenry (2000, 292). Hence, the research suggests that for climate change policies to be successful, they must be translated into local issues that are viewed as pressing and significant (Bulkeley 2000).

Some larger cities have been engaged in climate-change-reduction strategies over the last number of years and have attracted attention at the local level. The work of the C40 group of global cities is a prime example (Bouteligier 2013). However, what medium-sized cities are doing to combat economic and climate crises has received limited attention. As a group, such cities constitute a sizable and growing population base; indeed, approximately 26 percent of Canadians live in cities with populations of between 50,000 and 300,000, with a further 20 percent living in cities of between 300,000 and 900,000 people.[2] The level of green action at this scale provides an important indicator of how deeply climate-change-mitigation and green economy policies have penetrated into core policy realms.

RESEARCH, METHODOLOGY, AND APPROACH

Our research investigated the climate change and green economy strategies of medium-sized Canadian cities and the perceptions of actors involved at that level. The goal was to answer the following research questions: To what degree have cities adopted environmental policies and green economic development policies with the intention of reversing the effects of climate change, while improving economic performance and job growth? To what degree are such cities equipped to deal with climate change? Are cities and regional jurisdictions working together to mitigate climate change, and, if so, in what ways? Are medium-sized cities doing enough to combat climate change and promote green job growth? What unique challenges do mid-sized cities face in developing green growth plans?

A scoping exercise identified 27 mid-sized cities across Canada (characterized here as cities with populations of between 50,000 and 900,000). City websites were examined to determine whether a city had a green economic plan. Most mid-sized cities surveyed had some form of sustainability plan in place, but only three to six of these cities appeared to be combining climate change initiatives with active green economic development activities. We then conducted a series of not-for-attribution interviews in the summer of 2013 in three mid-sized Ontario cities. Municipal leaders in government, community, business, and labour were identified and interviewed; a total of eight extensive interviews were conducted with key informants. Neither cities nor respondents are identified, and interview material is synthesized and anonymized.

Before presenting the results of our analysis, however, it is important to offer the definition of a green economy used in our study. It followed the United Nations Environment Programme (UNEP), which defines a *green economy* as

> one whose growth in income and employment is driven by public and private investments that reduce carbon emissions and pollution, enhance energy and resource efficiency, and prevent the loss of biodiversity and ecosystem services. These investments need to be catalyzed and supported by targeted public expenditure, policy reforms and regulation changes. This development path should maintain, enhance and, where necessary, rebuild natural capital as a critical economic asset and source of public benefits,

especially for poor people whose livelihoods and security depend strongly on nature. (UNEP, n.d.)

In other words, a green economy is one that integrates policies that promote economic development and mitigate climate change, while improving the lives of people who live in the community.

All three mid-sized Ontario cities reviewed in this study have some degree of GHG emissions-reduction strategy as part of their overall environmental policy. They apply a variety of policy instruments, such as tax rebates, that act as incentives for existing businesses to green their products, services, processes, and infrastructure. They are also greening municipal buildings and fleets as well as conducting public waste, water, and energy awareness campaigns. While the extensiveness of a city's environmental and energy policies vary, these municipalities are demonstrating some commitment to combatting climate change.

But going green is not the same as greening the local economy. To be sure, a city can have a comprehensive GHG-emissions-reduction strategy, but have no plan to stimulate the growth of the local economy by promoting a growing green technology product or services sector. Indeed, despite increasing government incentives and the anticipated demand for green products and services, few Canadian mid-sized cities are aggressively pursuing a strategy to grow their local economy through green sector development in either their environmental or their economic strategies. Moreover, while cities often have environmental committees and economic development committees, only a few cities have fused their mandates to service both the environment and the economy.

Mid-sized cities may, however, come to adopt green economic policies when they recognize the co-benefits of doing so. For example, by investing in green energy to reduce municipal GHGs, a city may also attract new investors looking for a greener community. In fact, our mid-sized-city survey and interviews reveal that cities that have actively engaged in green economic development strategies have been motivated by the fact that they see the direct and immediate benefits of these measures to promote economic renewal in their communities.

Some cities go further by positioning themselves as local competition states (Cerny 2005; Dicken 1994) and are making strides to strengthen their local economies by investing in and developing a green technology or green energy sector. Their commitment to going green is not only to reduce GHG emissions, if indeed that is their primary motivation at all. Instead, these cities also focus on creating jobs and attracting investment by taking advantage of increasing opportunities in green energy and technology.

Furthermore, the literature identifies six pillars of a green economy: (1) renewable energy, (2) green buildings, (3) clean transportation, (4) water management, (5) waste management, and (6) land management

(Burkart 2009). Municipal and local governments have a direct hand in developing, implementing, and delivering services related to these six pillars, highlighting the enhanced role that municipalities should play in green economic development. These areas offer considerable potential for economic activity and employment opportunities. Moreover, given that public infrastructure, which is closely linked to cities, is aging and in need of upgrading and replacement, investment in these six areas will be substantive. Canada faces a $123 billion municipal infrastructure deficit; even now, some 80 percent of the current infrastructure is past its service life (Mirza 2007, 15). Coupled with the fact that population growth is concentrated in urban environments (Charbonneau, Ouellet, and Milan 2011), there is significant opportunity for cities to match urban-based economic development with green growth initiatives.

Like our division of the literature on the role of cities, we divide the results of our policy scan and interviews into two parts: those that present opportunities and those that present challenges for greening the local economy.[3]

MID-SIZED-CITY SCAN AND INTERVIEW FINDINGS

Part 1: Positive Green Economic Developments in Mid-sized Canadian Cities

Promoting Carbon-Friendly Employment: What Kind of Green Jobs?

Mid-sized cities in Ontario are growing green jobs in both the public and the private sectors. Compared with the rest of Canada, Ontario has the highest demand for green jobs; some 13 percent of these jobs are in green energy, and another 8 percent are in energy efficiency (Evergreen and ECO Canada 2012). Green energy jobs at the municipal level typically fit into one of several categories; the most prominent include manufacturing of renewable energy parts and components; retrofitting of existing buildings and new construction of energy-efficient buildings; installation of new technologies, such as solar panels; and engineering and planning. Many jobs do not require existing workers to be completely retrained. As one labour representative explained, "There are no new methods of turning a screw, but the context in which you are doing it may be dramatically different" (Interview, 16 July 2013).

Our interviews revealed that more green jobs are being created in the private sector. Many of these jobs, however, are temporary, such as those required in the construction of manufacturing facilities. Permanent jobs are often created in the manufacturing and installation of renewable energy technologies; in Ontario's mid-sized cities, these jobs have been most typically related to the solar energy industry. Indirect jobs in green technology and manufacturing include those in marketing, communications, and finance. Engineering has been cited as a field for which it is particularly difficult to find qualified workers in Ontario, and because it is typically a male-dominated field, one large Ontario manufacturer involved in renewable energy is actively promoting the engineering field among female youth.

These jobs, we are reminded, come in many shades of green. As a labour official explained,

> Working in a plant that makes a plant-based construction material that can replace steel, and has a total lower carbon footprint, and it doesn't entail mining and the whole carbon chain, is probably a greener job. So making the bolts for a wind tower, those are green jobs of varying shades of green. The greenest of green jobs would be doing energy conservation practice, upgrading homes, and in the process, or building energy-efficient homes, with low-carbon footprint through the whole chain, and instructing the owners or occupants of why you're doing it, and how they can continue to make it better, and continue to operate the home, so that it saves them money and keeps the carbon footprint down. (Interview, 16 July 2013)

The term *green job* has been widely used by municipal actors to categorize a range of jobs that have positive environmental impacts compared to other forms of employment. Since the meaning of *green job* continues to be debated, it makes sense that policy-makers would use the term to encourage citizens to buy in to new policies and programs. And it is clear that green jobs are growing in those mid-sized cities that are adapting their local environment to a low-carbon economy, while marketing their assets to attract green sector industries.

One city official made particular note that it was important to frame the discussion in a wide and inclusive fashion in order to win wide support for greening the economy and to help educate business, labour, community, and government participants about the economic as well as environmental benefits of promoting a low-carbon footprint when pursuing economic development.

> Let's describe the kind of economy we want. A low-carbon economy, a resilient economy, a diverse economy, a clean economy. We spend some time talking about that, and we use economic terms. And the lesson learned around using that term and defining its broader meaning creates a bigger tent that keeps everybody at the table. (Interview, 24 July 2013)

Policy Learning and Exchange

Policy learning and sharing best practices are cities' fundamental first steps in taking on an increasingly large policy role. Shifting away from hierarchical models of statecraft, sustainable economic policy development demands promoting horizontal modes of decision making that involve all levels of government and sectors of society (Forstater 2004, 2006; Betsill and Bulkeley 2004). In the shift from government to governance, we have witnessed a movement toward non-traditional models of governance, and this has allowed influential networks to develop, both formally and informally (Corfee-Morlot et al. 2009). These transnational municipal networks (TMNs) allow cities to develop creative and innovative policy solutions to complex problems, such as climate change and green growth strategies (Toly 2008). Promoted in regions such as the European Union, TMNs provide a forum and platform for global, horizontal policy learning (Betsill and Bulkeley 2006).

Our survey indicates that Ontario's mid-sized cities are learning from, and sharing best practices with, other cities in Ontario, North America, and Europe that are moving toward a low-carbon economy. Because of the high cost of natural gas in Europe, some cities on that continent have invested in research and development of district heating and cooling systems in urban centres. Ontario city representatives have attended European conferences to learn how such technologies might be adapted to the province's conditions. Some of Ontario's cities have also been approached by officials in cities as close as one town over or as far away as California. Sharing best practices may occur one to one – a city contacts another to learn about its programs – or in a network setting, where member organizations convene regularly to share knowledge or strengthen their overall marketability. For example, nine cities in the Ottawa, Niagara, and Windsor-Essex corridor belong to the Ontario Clean Technology Alliance, which focuses on promoting the region for new direct foreign investment in renewable energy. Rather than being in competition, cities in this group work together to attract investment to the region.

A city official provided a very concrete demonstration of the kind of networking that is emerging among mid-sized cities.

> We benchmarked our plan against European examples. So we look to Europe as a benchmark for where we wanted to go, and so that started an ongoing exchange with Europe as we moved into implementation. So right now I'm part of a project called the Transatlantic Urban Climate Dialogue.... It's a two-way street. Our learning is around implementation strategies, policies, techniques, institutions on the premise that Europe is ahead of us by decades. (Interview, 24 July 2013)

Such evidence suggests that municipalities that are leaders in greening cities are well on their way to strategically positioning themselves to become more effective players in green economy initiatives.

Taking Advantage of Federal and Provincial Green Funding Initiatives

Municipalities make active use of funding sources from upper levels of government to pursue environmentally friendly projects. The provinces make a number of sources available, especially those connected with municipal infrastructure initiatives, and they are a major source of funding for cities pursuing a green agenda. While the federal government has been reluctant to introduce policies that will combat climate change, it has provided some funds – although far more limited than those available from the provinces – for cities to stimulate their own economies using green initiatives. For example, the Green Municipal Fund is a federally funded program filtered through the FCM. The Canadian government has endowed $550 million for city-based projects in five sectors: brownfields, energy, waste, water, and transportation (Thompson and Joseph 2011). The FCM works with cities to help them tap into the funds, while providing guidance and knowledge.

Some cities have looked to other federal programs to help stimulate their local green economy. For example, FedNor provides funds for community economic development, innovation, and business growth and development in Northern Ontario communities. While organizations like FedNor do not explicitly focus on the green economy, some cities in our survey have used it to do so as part of their green economic diversification strategy.

The federal government's provision of green funding, given its otherwise hostile approach to regulating GHG emissions, might be explained as a case of using such funding to show that it is actually being proactive in promoting and investing in green economic development. Of course, such investment pales in comparison to the massive subsidization and promotion of conventional development – including dirty oil – that rests at the core of the Conservatives' national economic development strategy.

The introduction of austerity agendas by higher levels of government in the wake of the 2008 economic crisis has, however, meant that funding for municipalities has become more difficult to obtain. A key informant from the labour sector argued that the feds "are turning off the taps to grant money" (Interview, 16 July 2014), money that is crucial for start-up business initiatives in the green sector. Moreover, local government is often challenged by a lack of policy coherence at the federal and provincial levels. A municipal official observed that senior levels of government fail to agree on how best to deal with climate change. "We have a province that wants to do cap and trade and a federal government that doesn't know what they want to do" (Interview, 1 August 2013), a comment that tends to support the NRTEE analysis presented earlier. Such a contested policy environment poses significant challenges and dangers for cities attempting to navigate in such politically charged waters.

Part 2: Persistent Challenges and Opportunities

Public-Private Partnerships and Capacity

Mid-sized cities have partnered, or plan to partner, with the private sector in ways that can help clean up the local environment and/or increase green economic growth and jobs. Key informants cite a variety of reasons why municipalities and private industry are engaging in public-private partnerships (P3s) as one part of a policy toolkit that some of them believe can sometimes be usefully employed. Reasons why P3s can be attractive to cities include the problem of limited municipal resources, a shared lack of faith in the longevity of Ontario's GEA, and a conviction that job creation is not solely the responsibility of municipal government. Cities that have experienced large employment losses, an exodus of working-age people, and, as a result, a shrinking tax base, appear to be particularly in need of P3s to stimulate green industries and green jobs that will draw people back. Moreover, some city officials see the private sector as providing greater legitimacy to public sector projects, as one such official explained.

> Some of these programs need the private sector. The private sector adds some credibility as well. Not that you can't do it from the public sector, but it kind of makes sense for everyone. And I think it adds value to the project. And it also offers other opportunities to other companies to look at working with the public sector. It's an opportunity for growth. (Interview, 26 July 2013)

To be sure, most green jobs thus far have, in fact, been created by private business; hence, engaging with business is seen as key to making progress on green employment and environmentally sustainable practices. Some mid-sized cities, for example, are exploring opportunities for P3s through their local electricity utilities. Many local utility companies are wholly owned by the city they service, and some have both regulated and unregulated branches. While the regulated branch simply meets the energy needs of its customers, the unregulated arm is free to engage in commercial business, such as working with the private sector to build or grow the local energy infrastructure. It is these areas that present opportunities for P3s.

P3s are not a panacea, however, as there are numerous potential problems. For example, some cities are partnering with private companies to turn city waste into renewable energy, a process that is eligible for premium payments under Ontario's FIT program. While this arrangement can help solve a city's waste problems, it is probable that only the private partner will receive FIT payments. In other words, waste from goods purchased by municipal taxpayers may fuel profit for private industry but not bring complementary financial benefits to the city. If

P3s are to be a useful and effective policy tool, they must be framed in a manner that creates benefits for the wider local community; however, since they have been structured as mechanisms that shift public resources to private sector actors, their progressive potential is actually quite limited (Whiteside 2013).

Measures like P3s become attractive to cities in large measure because of the very limited policy and resource capacities open to the municipal level of government. A municipal official articulated the challenge this way. "Generally, we [municipalities] have no really powerful tools given to us by the government for attracting investment. You have to do it through sort of your own persuasive communication" (Interview, 24 July 2013). However, it is also the case that municipal capacity to deal effectively with the private partners involved in prospective P3s may be quite limited, especially when senior governments push this as a solution to the financial constraints in which cities find themselves (Whiteside 2013).

Level of Internal Organization and Leadership

Leadership from the mayor, city manager, and city council is a key ingredient in making the green economy flourish in mid-sized cities. Cities with strong plans have passionate and committed mayors, and these mayors surround themselves with staff and representatives from civil society who are also passionate about the green economy as a way to stimulate economic growth, job creation, and social inclusion. On the other hand, a decline in city commitment to green economy development often occurs when a mayor is elected and a city manager hired who are reluctant to recognize such potential. Toronto is a case in point, where an environmentally aware mayor, David Miller, was replaced by a very narrowly constituted, business-oriented Rob Ford. Ford's aggressive tax-reduction stance, and a political philosophy that allowed little place for government leadership to drive change, resulted in minimal public investment in climate change mitigation and green growth.

Our key informants characterized those cases of unsuccessful green economy strategies at the city level as "piecemeal" because they had varied levels of commitment from, and limited integration into, different city departments. Successful plans are not "mission statements" about why a city should be "greener"; rather, they are multi-year, intra-department strategic plans that are reviewed on a regular basis (by a committee, for example) and adapted, with changes, to the municipal environment and industry, provincial and federal policies and programs, and the wider global economy. Good strategic plans are concrete strategies that outline actual measures for drawing foreign investment to these cities. Moreover, successful city plans have a holistic understanding of sustainability. Sustainability is not only about protecting the local environment but also about social inclusion for citizens by ensuring job creation and affordable housing. It is clear that successful green strategies involve considerable planning, investment, political and community buy-in, and active leadership on the ground. In the fragmented social and political world that local government constitutes, this can be a challenge to achieve and maintain over time. The rewards, however, can be considerable.

Social Dialogue and Relationship with Local Non-governmental Organizations

Cities with green leadership and well-integrated green economy strategies tend to be more open to multi-stakeholder input than cities that are less engaged in green economic development. In these cities, local universities and colleges, business organizations, environmental groups, and labour are generally included at the discussion table in meaningful ways. However, not all organizations may support a city's green economy initiatives. Support can depend on how a city defines *green economy*. For example, hosting green industries does not necessarily mean a sustainable social and natural environment.

Some city officials view certain organizational interests as being unrealistic in their expectations. Consequently, they may draw a line between those whose input they deem to be valuable and doable and those whose input they view as utopian and obstructionist. These latter groups may make it difficult for cities to follow through on their green energy plans. For example, anti-wind groups have made it a challenge for some windier cities to issue permits for wind turbines within city limits. On the more positive front, while labour and environmental movements have historically been at odds, they are increasingly working together as "blue-green alliances" under a shared credence that a truly sustainable economy is one that is good for all people as well as the planet.

It is evident that while unanimity may not be necessary, broad involvement and support from local community actors is an essential ingredient for successful outcomes. A movement away from traditional hierarchical modes of governance requires recognition that climate change mitigation and green economy policy cannot solely be a top-down endeavour. The citizenry must be active and willing to partner in creating the move toward a greening economy (Katz and Bradley 2013). Civic participation can lead to creative and more responsive decision making, whereas the exclusion of citizens generally results in entrenched resistance and policy blockages (Corfee-Morlot et al. 2009, 63–65; Portney 2005; Thompson and Joseph 2011; Schreurs 2008).

Ontario's Green Energy Act and Municipal-Provincial Relations

While the GEA has been deemed to be a broadly progressive and useful set of policies for helping attract investment to Ontario, including foreign direct investment, in the renewable energy sector municipal representatives have also expressed disappointment and frustration with it because it further removes cities from participating in the development and planning of energy grids and programs. For example, one city official pointed to provincial discourse that called for building larger power plants as a way to meet increasing energy demands, despite the fact that many cities had already developed energy plans using community-based smart grids and conservation. This provincial limitation of power serves to only further restrain cities already limited under the *Municipal Act*, which minimizes cities' involvement in energy development and planning.

However, the largest city in the province has been exempt from many of these constraints. The *City of Toronto Act, 2006* bestowed on the city broader legislative and taxation control than other Ontario cities. As a result, some mid-sized cities believe that the provincial government is restricting and dismissing their role in energy and economic development. But the Ontario FIT program version 3.0 is moving toward a more community-based model, one in which small-scale projects are given priority over larger, corporate-owned energy projects. Thus, cities are slowly being included at the table in provincial energy grid planning.

Ontario's Feed-in-Tariff Program

While many environmental groups, businesses, and municipal governments agree that the FIT program has served as an incentive to attract green energy investments to Ontario's cities, some contend that it has also created its own challenges. The Ontario Power Authority approval process for FIT rates has often been long and unpredictable, something that some city officials and business community representatives argue has turned investors away from locating in the province altogether. Because energy policy can change with political leadership, this is an important factor in decision-making. For example, opposition Conservative leader Tim Hudak threatened during the Ontario election campaign in 2011 to scrap the FIT program, and some large foreign investors may have been reluctant to count on the GEA when making their investment decisions during this period. Indeed, one city official pointed out that a business deal had fallen through with one foreign investor – not because the city could not meet its needs, but because the volatility of the province's politics and policies made the investment too risky. On the other hand, since most large investors recognize that public policies can shift, the FIT program has not always been the driver for foreign investment. Instead, some cities believe that it is more what they as municipalities offer – rather than what the province is offering – that encourages a company to set up within city limits. Those companies in the renewable energy business already located in Ontario have apparently developed their own Plan B in case the FIT program is dissolved by future governments.

The local content requirement, which mandated that a certain percentage of labour and materials be sourced from Ontario to qualify for FIT rates, was scrapped in 2013 after a WTO ruling that it conflicted with international trade regulations. While labour organizations lamented the decision, some business organizations actually welcomed the change, arguing that it makes the green energy industry more competitive and thus has greater potential to create more stable employment over the longer term. As a business key informant commented, "The [XXX] network was never a fan of the restrictions placed on [it] by 'buy local.' We like open markets. And so we were glad to see that the government is going to be opening up competition" (Interview, 21 July 2013). The division of views within labour over the value of employing an activist state in pursuing a green jobs strategy points to an important area of ongoing policy contestation.

The case of the FIT program demonstrates very clearly the role that international trade agreements and bodies like the WTO have over domestic actors in their construction of green policies (McBride and Shields 2013; Sinclair and Trew, Ch. 1 this volume). Activist government policy at whatever level is vulnerable to the interests of international capital embedded in neoliberal trade regimes. This presents a formative obstacle to, and sets real limits on, activist government green-job-promotion policies.

CONCLUSION

Potentially, there are tremendous benefits to greening the economy and fostering carbon-friendly job growth at the municipal level. There are also, however, numerous barriers to local action. Cities interested in mitigating climate change and promoting green growth initiatives are often faced with financial obstacles, incomplete information, lack of authority to act, insufficient capacity challenges, lack of coordination with senior levels of government, a shifting public policy agenda and policy environment, and the ongoing challenge of building a set of diverse interests at the local level that constructs the kind of broad support that can move a green agenda forward (Robinson and Gore 2005; Schreurs 2008; Courchene 2007). Understanding progress at the level of mid-sized cities is important because they are substantive population centres in Canada, and they continue to grow. Unlike our largest cities, they also tend to have more limited resources, and, as smaller entities, their voices are often even more muted by larger governments and societal interests. Yet understanding what kind of municipal leadership and progress on green job strategies is occurring in mid-sized cities provides a kind of litmus test for how robust the progress on green policy development in Canada generally is.

Climate change and climate change mitigation are generally identified as national- and international-scale policy problems. But such a traditional and hierarchical understanding of climate change mitigation can impede local action and the necessary dialogue at the local, cross-jurisdictional, and societal levels necessary for effective action at the municipal level (Bai 2007; Gore 2010). Without enhanced cross-collaboration, climate change mitigation and green economic development will continue to be operationalized as nothing more than a global issue to be addressed exclusively by higher levels of government (Bulkeley 2011). Given that local governments often face budgetary and other financial obstacles to developing and implementing policies and programs, it is evident that economically sustainable environmental policies at the local level will also lead to coordination with the private and non-profit sectors.

While experts often discuss various frameworks for local government action on climate change and green economic development, the reality is that such frameworks are often not strengthened through legislative change. Municipalities in Canada, for example, have no formal constitutional status, and, with no national or provincial framework supporting their role in greening the economy, their ability to effect long-term, systemic change is inhibited. Having limited authority, cities are greatly constrained in the use of policy tools to, for example, implement revenue-raising mechanisms, such as user fees and property tax reform, to mitigate climate change and promote green job growth. Further disabling is the fact that cities face great public resistance to increased taxes and other municipal charges and fees (Courchene 2007; Thompson and Joseph 2011; Toly 2008). Municipalities need to be able to show the benefits of such green action for the quality of the life, work, and economy of local residents and businesses, and they must be able to demonstrate that they have the power and ability to act effectively on a green agenda. Such an agenda could be strengthened by federal and provincial policy frameworks that are, as we have noted, largely lacking.

Yet, for climate change policies to be successful, they must operate not just at the global, national, and regional levels; they must also be translated into local issues that are viewed as pressing and significant (Bulkeley 2000). The vast majority of people in Canada live in cities. Cities are the places from which most GHG emissions emanate; they are the places where most of us work and where jobs are created. Consequently, by necessity, cities need to be prime actors in climate change policy and green growth activities. Municipalities need an enhanced understanding of best green practices, local policy tools, and strategies for overcoming local barriers – as well as an activated and empowered civil society, including trade unions – working at the local level to push forward a green cities agenda. It is essential that environmentally sustainable policies at all levels, including the local, be complemented by policies that actively seek to overcome the jobs-versus-environment divide. It is at this level that much more work needs to be done.

NOTES

1. The authors would like to thank the research assistance of Olivia Cimo, Victor LaPierre, and Kristina Sannuto.
2. Authors' calculations based on census data (Statistics Canada 2011).
3. As mentioned earlier, eight interviews were conducted in the summer of 2013 in the three cities with key informants from the municipal government, labour, business, and non-governmental constituencies. All interviews were conducted in confidentiality, and the names of the interviewees are withheld by mutual agreement. In addition, the quotations that follow do not identify the cities that the respondents represent, but they do indicate their constituency.

REFERENCES

Aulakh, R. 2013. "UN Climate Talks: Japan, Australia Join Canada as 'Villains.'" *Thestar.com*, 22 November. http://www.thestar.com/news/world/2013/11/22/un_climate_talks_japan_australia_join_canada_as_villains.html.

Bai, X. 2007. "Integrating Global Environmental Concerns into Urban Management: The Scale and Readiness Arguments." *Journal of Industrial Ecology* 11 (2):15–29.

Betsill, M., and H. Bulkeley. 2004. "Transnational Networks and Global Environmental Governance: The Cities for Climate Protection Program." *International Studies Quarterly* 48 (2):471–93.

—. 2006. "Cities and the Multilevel Governance of Global Climate Change." *Global Governance* 12 (2):141–59.

Bouteligier, S. 2013. *Cities, Networks, and Global Environmental Governance: Spaces of Innovation, Place of Leadership*. New York: Routledge.

Bradford, N. 2002. *Why Cities Matter: Policy Research Perspectives for Canada*. CPRN Discussion Paper No. F23. Ottawa: Canadian Policy Research Networks.

Brenner, N., and N. Theodore. 2003. *Spaces of Neoliberalism: Urban Restructuring in North America and Western Europe*. New York: Wiley.

Bulkeley, H. 2000. "Down to Earth: Local Government and Greenhouse Policy in Australia." *Australian Geographer* 31 (3):289–308.

—. 2011. "Cities and Subnational Governments." In *Oxford Handbook of Climate Change and Society*, ed. J.S. Dryzek, R.B. Norgaard, and D. Schlosberg, 464–79. New York: Oxford University Press.

Burck, J., F. Marten, and C. Bals. 2013. *The Climate Change Performance Index: Results 2014*. Bonn: Germanwatch / Climate Action Network. https://germanwatch.org/en/download/8599.pdf.

Burkart, K. 2009. "How Do You Define the 'Green' Economy?" *Mother Nature Network* (blog). 9 January. http://www.mnn.com/green-tech/research-innovations/blogs/how-do-you-define-the-green-economy.

Calamai, P. 2007. "The Struggle over Canada's Role in the Post-Kyoto World." In *Innovation, Science, Environment: Canadian Policies and Performance, 2007–2008*, ed. G. Bruce Doern, 32–54. Montreal and Kingston: McGill-Queen's University Press.

Cerny, P.G. 2005. "Political Globalization and the Competition State." In *Political Economy and the Changing Global Order*, 3rd ed., ed. R. Stubbs and G.R.D. Underhill, 300–09. Don Mills, ON: Oxford University Press.

Charbonneau, P., G. Ouellet, and A. Milan. 2011. "Population Growth: Subprovincial, 2010." Last modified 20 July. http://www.statcan.gc.ca/pub/91-209-x/2011001/article/11510-eng.htm.

Corfee-Morlot, J., L. Kamal-Chaoui, M.G. Donovan, I. Cochran, A. Robert, and P.J. Teasdale. 2009. "Cities, Climate Change and Multilevel Governance." OECD Environment Working Papers No. 14. Paris: OECD Publishing.

Courchene, T. 2007. "Global Futures for Canadian Global Cities." *IRPP Policy Matters* 8 (2):1–36.

David Suzuki Foundation. 2012. *All over the Map: A Comparison of Provincial Climate Change Plans*. Vancouver: David Suzuki Foundation. http://www.davidsuzuki.org/publications/downloads/2012/All%20Over%20the%20Map%202012.pdf.

Demerse, C. 2011. *Reducing Pollution, Creating Jobs: The Effects of Climate Policies on Employment*. Drayton Valley, AB: Pembina Institute. http://www.pembina.org/reports/reducing-pollution-creating-jobs.pdf.

Dicken, P. 1994. "Global-Local Tensions: Firms and States in the Global Space-Economy." *Economic Geography* 70 (2):101–28.

Evergreen and ECO Canada. 2012. *The Green Jobs Map: Supplementary Ontario Report*. Labour Market Research Study. Toronto and Calgary: Evergreen / ECO Canada. www.eco.ca/ecoreports/pdf/Green-Jobs-Map-Ontario-2012.pdf.

Forstater, M. 2004. "Green Jobs: Addressing the Critical Issues Surrounding the Environment, Workplace and Employment." *International Journal of Environment, Workplace and Employment* 1 (1):53–61.

—. 2006. "Green Jobs: Public Service Employment and Environmental Sustainability." *Challenge* 49 (4):58–72.

Gore, C. 2010. "The Limits and Opportunities of Networks: Municipalities and Canadian Climate Change Policy." *Review of Policy Research* 27 (1):27–46.

Katz, B., and J. Bradley. 2013. *The Metropolitan Revolution: How Cities and Metros Are Fixing Our Broken Politics and Fragile Economy*. Washington, DC: Brookings Institution Press.

Kousky, C., and S. Schneider. 2003. "Global Climate Policy: Will Cities Lead the Way?" *Climate Policy* 3 (4):359–72.

Leitner, H., J. Peck, and E.S. Sheppard, eds. 2007. *Contesting Neoliberalism: Urban Frontiers*. New York: Guilford.

Maclean's. 2013. "Canada Ranks Worst on Climate Policy among Industrialized Countries: Report." Canadian Press, 18 November. http://www2.macleans.ca/2013/11/18/canada-ranks-worst-in-developed-world-on-climate-policy-european-report/.

Mahon, R., C. Andrew, and R. Johnson. 2007. "Policy Analysis in an Era of Globalisation: Capturing Spatial Dimensions and Scalar Strategies." In *Critical Policy Studies*, ed. M. Orsini and M. Smith, 41–64. Vancouver: University of British Columbia Press.

McBride, S., and J. Shields. 2013. "International Trade Agreements and the Ontario Green Energy Act: Opportunities and Obstacles." In *Climate@Work*, ed. C. Lipsig-Mummé, 41–56. Halifax: Fernwood Publishing.

Miller, D. 2012. "Achieving Coexistence: Integrating the Environment and the Economy." Keynote speaker at the Third Annual EnSciMan Research Seminar, Ryerson University, Toronto, 13 March.

Mirza, S. 2007. *Danger Ahead: The Coming Collapse of Canada's Municipal Infrastructure*. A Report for the Federation of Canadian Municipalities. Ottawa: Federation of Canadian Municipalities. November. https://www.fcm.ca/Documents/reports/Danger_Ahead_The_coming_collapse_of_Canadas_municipal_infrastructure_EN.pdf.

NRTEE (National Round Table on the Environment and the Economy). 2012. *Reality Check: The State of Climate Progress in Canada*. Ottawa: NRTEE.

Portney, K. 2005. "Civic Engagement and Sustainable Cities in the United States." *Public Administrative Review* 65 (5):579–91.

Robinson, P.J., and C.D. Gore. 2005. "Barriers to Canadian Municipal Response to Climate Change." *Canadian Journal of Urban Research* 14 (1):102–20.

Schreurs, M. 2008. "From the Bottom Up: Local and Subnational Climate Change Politics." *Journal of Environment and Development* 17 (4):343–55.

Statistics Canada. 2011. "2011 Census." http://www12.statcan.gc.ca/census-recensement/index-eng.cfm.

Thompson, D., and S.A. Joseph. 2011. *Building Canada's Green Economy: The Municipal Role*. Ottawa: Federation of Canadian Municipalities. http://www.fcm.ca/Documents/reports/Building_Canadas_green_economy_the_municipal_role_EN.pdf.

Toly, N. 2008. "Transnational Municipal Networks in Climate Politics: From Global Governance to Global Politics." *Globalizations* 5 (3):341–56.

UNEP (United Nations Environment Programme). n.d. "Green Economy: About GEI – What Is the 'Green Economy'?" http://www.unep.org/greeneconomy/AboutGEI/WhatisGEI/tabid/29784/Default.aspx.

Whiteside, H. 2013. "P3s, Austerity, and Canadian Public Procurement: Sustainable Development or Sustainable Risk?" Presented at the W3 International Conference 2013, "Work in a Warming World: Labour, Climate Change and Social Struggle," Toronto, 29 November–1 December.

CHAPTER 10

RENEWABLE ENERGY, SUSTAINABLE JOBS: THE CASE OF THE KINGSTON, ONTARIO, REGION

MEGAN MACCALLUM, LINDSAY NAPIER,
JOHN HOLMES, AND WARREN EDWARD MABEE

As part of a strategy to become Canada's most sustainable city, Kingston, Ontario, has supported renewable energy–generation projects in and around the municipality. An initial review of the region conducted in 2011 indicated that Ontario's *Green Energy and Green Economy Act* (GEA), combined with the presence of wind and solar resources, had served to initiate a cluster of business activity. This chapter explores the definition of sustainable green jobs, proposes categories of green jobs, and highlights issues associated with developing a common definition. It also considers the use of the term *green jobs* in the international and Canadian contexts. It then categorizes green employment related to the renewable energy sector around Kingston and addresses how we can track progress in each category. Finally, it assesses the role of renewable energy as a driver for employment in the Kingston region.

INTRODUCTION

As the 21st century unfolds, one of the most fundamental challenges before global society is the need to balance environmental sustainability and economic prosperity. Rising greenhouse gas (GHG) emissions, largely associated with fossil energy use, and the potential for catastrophic climate change as a result of these emissions, have highlighted the need for clean and renewable energy sources (IPCC 2013).

Work in a Warming World, edited by Carla Lipsig-Mummé and Stephen McBride. Kingston: School of Policy Studies, Queen's University.

Canadians are becoming increasingly aware of their vulnerability to climate change. Under the Copenhagen Accord, the federal government committed to a 17 percent reduction in GHG emissions below 2005 levels by 2020 (UNFCCC 2009). The most recent reports indicate that the current trend in Canadian GHG emissions will result in total emissions of about 734 megatonnes (MT) per year in 2020, about 122 MT above the national target (Environment Canada 2014). In 2009, approximately 90 percent of Canada's GHG emissions came from the production or use of fossil energy (Environment Canada 2011). While Canada's energy mix is already fairly green – about 19 percent of total primary energy supply is renewable, with the greatest amount coming from hydroelectric sources (IEA 2014) – there is clearly an impetus for Canadians to consider increasing the amount of renewable energy in use in order to meet the Copenhagen goals.

In developing renewable energy, Canada's provincial governments have taken a strong lead, actively developing green policies in their respective regions. Examples of provincial strategies to increase renewable energy use include British Columbia's Energy Plan (BC MEMPR 2009), which sets provincial targets of 90 percent renewable electricity generation and zero net GHG emissions associated with thermal energy production. Alberta's current energy strategy includes language about increasing renewable production but also focuses on developing cleaner options for fossil energy (Alberta 2009). In autumn 2013, Quebec began a public consultation process on the future of the province's energy sector, including reducing GHG emissions from fossil fuel use and increasing the use of renewables in a variety of sectors (Mousseau 2013).

In Ontario, the Long-Term Energy Plan, published in 2013, and the GEA of 2009 set goals for renewable energy production and provided specific tools to implement new electricity generation capacity (Ontario Ministry of Energy 2013; Ontario 2009). The main tool created through the GEA was a feed-in tariff (FIT), which offered set rates over extended periods for electricity produced from various renewable sources (Mabee, Mannion, and Carpenter 2012). The GEA also specified certain proportions of domestic content (Ontario-made parts and labour) in order to drive employment in the province. From 2009 to 2013, the FIT program was divided into two programs: FIT – for projects greater than 10 kilowatts (kW) installed capacity – and MicroFIT – for projects less than 10 kW. In 2013, the World Trade Organization (WTO) ruled that the domestic content requirements were too stringent, given existing trade agreements; the implications of these changes are described in more detail later in this chapter. In 2014, a new competitive procurement program was introduced for projects greater than 500 kW, leaving the FIT program to service projects between 10 and 500 kW (Ontario Ministry of Energy 2013). To date, there have been three rounds of the FIT program, and projects have been announced for the first two rounds (OPA 2014c). As the original FIT program has evolved, one modification has been made:

the introduction of a priority points scheme to help determine projects for funding. Up to nine priority points can be awarded to a project based on project readiness, community participation, Aboriginal participation, municipal support, and electricity system benefits (OPA 2014b).

An interesting component of new energy policies is the expectation of increased employment. Some provincial documents use specific language to describe the sustainable green jobs associated with renewable energy production (BC MEMPR 2009). Other provinces, such as Alberta, recognize that the energy sector provides employment but do not identify green jobs as a specific target (Alberta 2009). Quebec's strategic goals include a more sustainable economy, which by definition must include employment (Mousseau 2013). In Ontario, the government claimed that energy projects approved under the GEA would generate 50,000 green jobs in the first three years (Ontario Ministry of Energy and Infrastructure 2011); no revised employment target has been set since the WTO ruled against parts of the GEA in 2013.

Green jobs associated with renewable energy production have been identified as an important component of local economies. The city of Kingston, Ontario – located at the eastern end of Lake Ontario, midway between Toronto and Montreal – is one example of a municipality that expects to benefit from the development of renewable power, both by improved environmental performance and by increased employment (Focus Kingston 2010). The implementation of the GEA has led to a number of new renewable energy projects in the region around Kingston. This chapter explores the relationship between renewable energy policy – in this case, the GEA – and sustainable employment in Kingston and the surrounding region.

DEFINING SUSTAINABLE GREEN JOBS

In assessing the impact of the GEA on employment in the Kingston region, one must first address the issue of definitions. There is considerable variability in the understanding of a sustainable green job. In its earliest use, authors tended to restrict *green jobs* to a limited number of fields or vocations (OECD 2012). The definition has evolved over time, but no consistent definition has been set. A recent report from the United States highlights some of the issues involved in creating a cohesive definition (Gülen 2011).

Key Components of a Definition of *Green Job*

One issue is that sustainable jobs need not be new jobs (Gülen 2011). While new jobs are often desirable (particularly from a government perspective), retained jobs are important because the greening of employment is part of the process of meeting increasingly stringent requirements for lower GHG emissions and increased energy efficiency (D. Parsons & Associates 2009). There is an emerging imperative to green both products and production processes, which almost inevitably means greening employment. In the European Union (EU), continued development of an emissions-trading scheme in conjunction with increasingly stringent targets for reducing GHG emissions has impacted many supply chains, creating new markets for fuels like wood pellets (Aguilar et al. 2012) but potentially limiting the markets for other Canadian products, such as oil sands outputs (Lapointe 2013). Similarly, the United States has indicated that environmental performance is a factor in the impending decision on the Keystone XL project (Cryderman 2014). Retaining jobs in existing sectors through a greening process, by implementing skills and shifting the focus to consider environmental considerations, could have a much greater impact than focusing on creating new jobs (Manitoba Education 2012). In fact, the United Nations Environmental Programme (UNEP) predicts that most workplaces will see modest changes to everyday occupational activities, which will improve overall environmental performance (Worldwatch Institute 2008).

A second issue is the suggestion that green jobs should incorporate a dimension of economic sustainability – that one might expect these jobs to continue to exist over an extended period – an important aspect that is often paid less attention when defining green jobs. It is known that a large number of new green jobs are temporary jobs associated with construction and start-up of new, greener processes (such as renewable energy projects). These jobs may be reported in a way that makes it appear as though green construction jobs are secure and rising (Gülen 2011). Simply reporting the number of green jobs created may fail to consider their permanence, along with other aspects of job quality such as wages, benefits, union representation, and the ability to move up the occupational ladder – all of which are important components of long-term employment sustainability (Sommers 2013). A shift to a green economy will result in many transitional jobs – temporary employment associated with construction and project implementation that will be relatively short-lived.

A third issue highlighted by Gülen (2011) is full-time versus part-time employment. Consider that green jobs are often classified by sector; thus, jobs in renewable energy, agriculture, or forestry, for example, may be included in an assessment of green employment (OECD 2012). In any sector, jobs may be direct (i.e., 100 percent related to the sector) or indirect (or induced) (i.e., part-time, service-oriented, or engaged in an element of a supply chain). In Canada's forestry sector, indirect jobs account for almost one-third of total employment (Natural Resources Canada 2014b, 45). Thus, estimates of green employment that include broad sector categorizations should track the proportion of direct and indirect jobs included.

Given the discussion above, it is possible to identify at least five categories of green jobs, as shown in Figure 10.1 below. The availability of data will obviously dictate whether it is possible to track employment in each of these categories, but each represents a distinct and important component of an emerging green economy.

FIGURE 10.1
Categories of Green Jobs

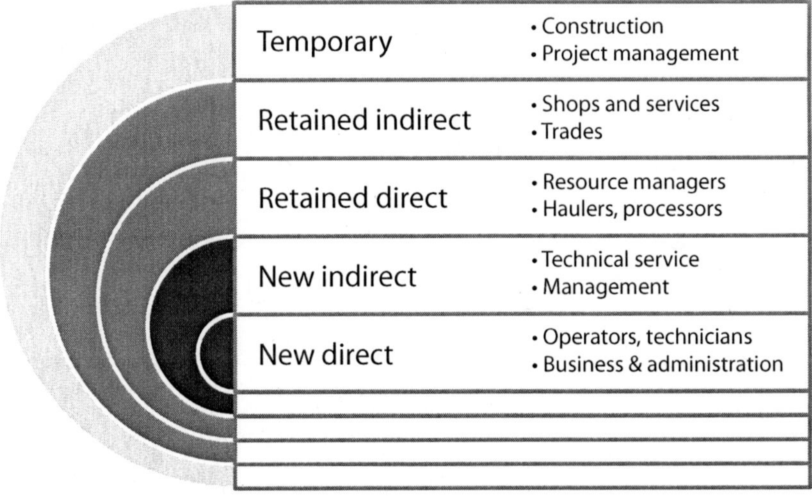

Source: Earley and Mabee (2011).

One final issue about the definition of sustainable green jobs is the nebulous nature of these terms. Indeed, activities in almost every sector of the economy can be considered green (Katz 2012). Furthermore, technologies considered green by some metrics (for instance, nuclear power viewed solely through the lens of GHG emissions) may not meet the criteria of sustainability when other environmental aspects are considered (for example, the disposal of spent nuclear fuels). The application of poor practices in forestry or agriculture, which are typically considered green industries, have potentially significant, negative environmental ramifications (Gülen 2011). Thus, the definition of green jobs should ultimately incorporate metrics (preferably universally adopted) to define the degree to which they might warrant the label.

Definitions of *Green Job* Currently in Use

Because there is such a wide range in what qualifies as a green job, it is helpful to track the development of this term and its current use in order to understand how we might better define green jobs in Ontario and particularly around Kingston.

One of the earlier definitions of green jobs dates back to 1999, when the Organisation for Economic Co-operation and Development (OECD) and Eurostat suggested that green jobs are associated with industry that produces environmentally friendly goods and services; by this narrow definition, only 2 percent of total employment across the EU could be considered green as of 2010 (OECD 2012). Similarly, the US Department of Commerce has created a definition for green jobs that focuses on performance metrics associated with GHG emissions and energy efficiency. The department estimated that the US green economy represented 1 to 2 percent of all economic activity in the United States in 2007 (US Department of Commerce 2010a, 2010b).

These definitions may be seen to be limited, however, as they do not capture employment that supports a green agenda in more conventional companies and also because they do not provide any quantitative means to compare workers in different sectors. Thus, conventional employment that is potentially sustainable (for example, jobs in agriculture or forestry) cannot be easily compared to conventional jobs in fossil resources, which by definition are unsustainable in the long term. By 2008, UNEP had provided a broader definition of green jobs, which included work in agriculture as well as manufacturing, research, administration, and services and which could be seen as substantially preserving or restoring environmental quality (Worldwatch Institute 2008). More recently, the US Bureau of Labor Statistics has published two distinct definitions; they differentiate between jobs in companies that produce goods or services of benefit to the environment and jobs whose descriptions include improving the impact that a company has on the environment (Sommers 2013). The former definition encompasses most renewable energy companies, whose business plan involves providing a service that has a lower environmental footprint (as measured using specific indicators such as GHG emissions) when compared to conventional alternatives. A job under the second definition might include that of a sustainability manager brought in to improve the operations of an existing business without necessarily changing the goods or services it provides.

A 2011 review by the International Institute for Labour Studies pointed out that the lack of a standard definition for green jobs makes it difficult for governments to compare the efficacy of their policies and that there is a need to standardize working definitions to better inform employment policy (IILS 2011). This report makes an attempt to define green jobs in a systematic fashion, including recognizing jobs that are either created or maintained as a result of a shift toward a green economy – essentially acknowledging that jobs may be retained as industries improve their performance or that new jobs may be created as new industrial activity emerges (ibid.).

Canada has seen a number of definitions of green jobs emerge in different jurisdictions. At the federal level, the definition seems to be rooted in sectoral employment, particularly employment related to renewable energy, forestry, and agriculture; a recent federal report indicates that green jobs accounted for 11 percent of job growth across Canada between March and May 2012 (Evergreen and ECO Canada 2012). The Environmental Careers Organization (ECO Canada), which bills itself as the largest online resource in the country for environmental jobs, training, and certification, defines green jobs by activity and by training; any job that works to reduce environmental impacts directly and requires specialized skills or experience can be classified as a green job (ECO Canada 2010). The Globe Foundation in British Columbia has built on both ECO Canada's definition and the UNEP definition of green jobs, focusing on any activity that is able to reduce environmental impacts (Globe Foundation 2010). In a joint report authored for Toronto and the Peel and York regions in Ontario, the need to reduce environmental impacts is included in the green job definition, but the authors also suggest that a livable wage, one that includes benefits that can support upward mobility, is an important aspect of green jobs (D. Parsons & Associates 2009).

It can be seen from this discussion that definitions of green jobs are rarely comprehensive; no single definition described here addresses all of the issues identified in the previous section, and certain elements (such as the need for a living wage or long-term employment) are not explicitly addressed. Definitions within a single country can vary from agency to agency (as seen in the US example) or from government to government (as seen in the Canadian example).

In Ontario, the term *green jobs* has been used in reference to the GEA. As stated earlier, in 2009, the Ontario government asserted that the GEA would create up to 50,000 jobs associated with renewable energy development over a three-year period (Ontario Ministry of Energy and Infrastructure 2011). This statement might cause one to suspect that the definition of green jobs was left deliberately obscure; the GEA does not define green jobs, and thus no criteria can be applied to determine whether the objective has been successfully reached. This means that evaluating Ontario's success in creating green jobs through the GEA is difficult and that the opportunity for the government to defend the GEA on the basis of new employment generated is lost. Furthermore, the ruling of the WTO in 2013 has weakened the ability of the GEA to create employment in the province, and since that ruling, no revised target for green jobs has been set.

RENEWABLE ENERGY IN THE KINGSTON REGION

The city of Kingston has publicly set a goal of becoming Canada's most sustainable city, and, as part of this goal, it has developed a platform that incorporates cultural vitality, economic health, environmental responsibility, and social equity. This platform specifically refers to environmental issues such as climate change but also economic issues such as regional development and new job creation. One of Kingston's sustainability goals is to generate enough renewable energy to meet local needs (Focus Kingston 2010).

Kingston and the surrounding region is well situated to take advantage of renewable energy opportunities as the local geography is endowed with renewable resources, including biomass, sunlight, and wind. In addition, the city is home to a range of people and organizations with the necessary skill sets to make renewable energy happen. One such organization is SWITCH Ontario, a not-for-profit organization launched in 2002 to promote the development and implementation of energy-efficient and renewable energy technologies (SWITCH n.d.). With a membership of almost 100 companies and individuals, this organization has the capacity to support renewable energy projects across the region.

A number of local educational institutions are engaged in training highly qualified personnel able to work in the emerging renewable energy sector. For example, both Queen's University and the Royal Military College of Canada have introduced new programs to build research expertise in sustainability and energy systems. St. Lawrence College offers three programs in the renewable energy sector, focusing on energy systems engineering, wind turbines, and geothermal energy (St. Lawrence College n.d.). Graduates from programs across higher-level educational facilities have supported the development of business and services in Kingston related to the renewable energy sector (Napier 2011).

Biomass Energy

Biomass-based renewable energy can be generated using a number of pathways (Sims et al. 2010). Solid biomass from forest or agriculture can be combusted, thereby producing heat and/or electricity, or it can be converted, using biological or thermochemical means, to liquid fuels for transport. Biogas can be collected through anaerobic digestion of landfill material or farm manure (Mabee and Mirck 2011). Each iteration of Ontario's FIT program has recognized biomass-based electricity, electricity from biogas on-farm, and electricity from landfill gas; the current FIT program (which applies only to projects between 10 and 500 kW) offers base rates of $0.13, $0.077, and $0.164 per kilowatt hour (kWh), respectively, for these bioenergy options (OPA 2014c). Because these rates do not attach value to the potential heat output of a combined heat-and-power facility, they are too low to drive significant investment in these power options (Moore, Durant, and Mabee 2013). Indeed, in the first two rounds of FIT funding (2010 and 2011), only 2.8 percent of all contracts offered (48.6 MW) went to the three biomass options, and in the immediate Kingston region, only one on-farm biogas facility was approved: the 0.36 MW De Bruin Farms biogas facility on Wolfe Island (OPA 2014c).

While few biomass-to-energy projects were created under the GEA across southern Ontario, it should be pointed out that there are rich biomass resources in the Kingston area. One study suggests that as much as 260,000 dry tonnes of residues are produced every year from sawmills in eastern Ontario, which in turn could be used for bioenergy production (Levin, Krigstin, and Wetzel 2011). This figure is likely matched by unused annual growth in the forest (i.e., biomass that could be sustainably harvested, but is not currently used by the forest products sector) (Mabee and Mirck 2011). Agricultural biomass supply, including potential energy sources such as corn stover, wheat straw, and hay, is largely detailed through the census of agriculture. Recent estimates suggest that agricultural biomass availability reaches as much as 725,000 dry tonnes per year across the five predominantly agricultural counties to the east of Kingston; taken together, then, forest and agricultural biomass could provide about 10 percent of energy requirements in eastern Ontario (Mabee and Mirck 2011). In addition to the lack of proper incentive under the FIT program, a lack of good data on biomass availability in the region may contribute to the lack of uptake of this option (Calvert, Luciani, and Mabee 2014). Given the proper incentive and better data on resource availability, biomass energy options could become more important to the Kingston region in the future.

Wind Energy

Kingston's geographical location at the eastern end of Lake Ontario gives it ideal wind potential, particularly given the prevailing westerly winds, which gain speed as they travel across the expanse of the lake. The mean wind speed across most of Ontario, measured at 50 metres above ground, ranges between four and six metres per second; at the eastern end of Lake Ontario, wind speeds are routinely in the range of eight metres per second, making this area among the most suitable for wind farms in the province (Environment Canada 2008). Under the current (post-2013) FIT program, wind-based electricity receives a base rate of $0.115 per kWh. Approximately 34.9 percent of all contracts (615 MW) offered under the first two rounds of the FIT program went to wind-based electricity (OPA 2014c).

In the Kingston region, the GEA has resulted in the development of four wind-to-electricity project proposals since 2009, as detailed in Table 10.1 below. Only one of these projects has been approved to date (the Ernestown Wind Park, which is about to begin construction). Two offshore wind projects (Trillium and Windstream) were proposed early in the FIT program, before a moratorium on offshore wind was put in place (OPA 2014c). The existing wind farm on Wolfe Island predates the GEA but is included here for the sake of completeness.

TABLE 10.1
Wind Energy Projects in the Kingston Region as of 2013

Project name and information	*MW*	*Status*
Wolfe Island Wind Farm: TransAlta operates a wind farm consisting of 86 turbines on Wolfe Island	198	Complete*
Ernestown Wind Park: Ernestown Windpark Inc. (operating as Ernestown Windpark Limited Partnership) is a co-operative project consisting of five turbines	10	Approved
Amherst Island Wind Farm Project: Windlectric Inc. plans to construct 36 turbines	75	Under review
Trillium Power Wind 1: Proposes to engineer offshore wind farms in Lake Ontario	420	Proposed/ on hold
Windstream Wolfe Island Shoals Inc.: Proposes to develop an offshore wind farm	300	Proposed/ on hold

*This project was contracted under the Renewable Energy Standard Offer Program.
Source: OPA (2014c).

Solar Energy

Eastern Ontario has some of the greatest solar photovoltaic (PV) potential in Canada, and it is recognized as one of the world's sunniest places, falling just short of Rio de Janeiro, Brazil. Kingston has a yearly solar PV potential of 1,197 kWh per kW, ranking fifth among municipalities in Canada with the greatest amount of solar PV potential and tenth around the world (Natural Resources Canada 2014). Solar potential in the Kingston region ranges between 3.52 and 3.96 kWh per square metre of irradiation per day (Nguyen and Pearce 2010).

A 2010 study estimated the potential for solar PV around Kingston by considering a wide array of factors, including soil and land use classifications, digital elevation models, slope, aspect, latitude, albedo, clear sky index, and irradiation. Over 375,000 hectares (ha) across the region were found to be suitable for solar farm development, suggesting a theoretical regional capacity of about 90 gigawatts (GW) (Nguyen and Pearce 2010). A similar study examined rooftop availability over the same region, establishing that 5.7 GW of additional solar PV capacity could be achieved by installing panels on appropriate roofs (Wiginton, Nguyen, and Pearce 2010). The potential to expand solar PV in and around Kingston is dramatic.

Under the existing FIT program, contracts for all ground-mounted solar PV projects are currently offered at $0.291 per kWh, while rooftop projects greater than 10 kW and less than 500 kW receive between $0.329

and $0.346 per kWh (OPA 2014c). As such, solar PV projects receive the greatest incentive of all renewable energy types under the current FIT program and, not surprisingly, constituted the largest percentage of projects contracted under the first two rounds of the FIT program. Approximately 51.4 percent of projects (907 MW) contracted in the first two rounds were ground-mounted solar PV projects, while only 1.5 MW of projects used roof-mounted solar PV.

In the Kingston region, the GEA has resulted in six major solar projects under the FIT program, and they are currently under development. Two additional projects, already operating, were contracted under the Renewable Energy Standard Offer Program, which predated the GEA. In addition, a few MicroFIT applications (10 kW of installed capacity or less) have been received from Kingston and the surrounding region, totalling less than 1 MW of installed capacity (OPA 2014c). Only FIT-scale projects are included in Table 10.2 below.

TABLE 10.2
Solar PV Energy Projects in the Kingston Region as of 2013

Project name and information	*MW*	*Status*
First Light I: SkyPower/SunEdison built a ground-mounted solar PV installation in Stone Mills	9.1	Complete*
First Light II: SkyPower built a ground-mounted solar PV installation in Stone Mills	10.5	Complete*
Kingston Gardiner TS Odessa Solar Power Project: Axio Power Canada Inc./SunEdison Canada will construct a ground-mounted solar PV project near Odessa (located just west of Kingston)	10	Approved
Kingston Gardiner Highway 2 South Solar Project: Axio Power Canada Inc./SunEdison Canada will construct a ground-mounted solar PV project near Odessa	10	Approved
Kingston Gardiner TS Unity Road Solar Project: Axio Power Canada Inc./ SunEdison Canada will construct a ground-mounted solar PV project in Kingston	10	Approved
Little Creek Solar Project: Canadian Solar will construct a ground-mounted solar PV project in Lennox and Addington County	10	Approved
Napanee TS Taylor Kidd Solar Energy Project: Axio Power Canada Inc./ SunEdison Canada will construct a ground-mounted solar PV project near Millhaven (located west of Kingston)	10	Approved
Samsung Sol-luce: Samsung will build and operate a solar power cluster in the city of Kingston and Loyalist Township	100	Approved

*These projects were contracted under the Renewable Energy Standard Offer Program.
Source: OPA (2014c).

RENEWABLE ENERGY AND GREEN JOBS IN KINGSTON

As shown in Table 10.1 above, the largest wind farm in the region predates the GEA. In 2008, Canadian Hydro Developers, Inc. developed an 86-turbine wind farm on Wolfe Island, directly south of Kingston. The nameplate capacity of this facility is 198 MW of power, enough to power 75,000 homes in the region (TransAlta 2014). Development of the farm, which took over a year, employed over 400 construction workers at peak staffing, including trades, truck, and barge operators and other affiliated workers; at the current time, there are 12 permanent, new, direct jobs associated with the project (O'Meara 2013). A simple multiplier can be developed to predict new direct jobs per MW (12 per 198 MW, or 0.06 per MW) as well as the number of temporary jobs (400 per 198 MW, or 2.02 per MW). The multipliers are applicable to onshore wind developments, but, for the purpose of this study, we applied it to the two offshore wind farms currently on hold. In reality, offshore wind developments would probably require more temporary workers and perhaps more permanent staff to maintain their operations.

As shown in Table 10.2 above, the Kingston region is home to two operating solar farms. The first commercial solar farm in Canada, the First Light I facility, is a 9.1 MW farm constructed in the small rural municipality of Stone Mills, located about 20 minutes west of Kingston, by SkyPower and SunEdison (SunEdison 2014). A second solar farm, First Light II, is a 10.5 MW facility built close to the original facility (SkyPower Global 2014). The First Light I solar farm occupies approximately 36 ha of land and uses over 120,000 thin-film PV solar panels, generating enough power through the season to supply about 1,000 homes per year (SunEdison 2014). The project employed around 100 people during the construction phase, and two to four permanent jobs are associated with the facility (SunEdison 2014). It should be noted that for subsequent solar installations carried out by this company, estimates of permanent employment per unit of output drop significantly, to between one and two permanent employees for a 10 MW facility (SunEdison 2011). Again, simple multipliers can be developed for new direct jobs (2 per 10 MW, or 0.20 per MW, using the more conservative employment estimates) and temporary employment (100 per 9.1 MW, or 10.99 per MW).

This discussion of new direct and temporary employment addresses two of the five green job categories identified in Figure 10.1 above. Questions remain, however, about indirect employment (both new and retained) and retained direct employment. In the wind energy sector, indirect employment might include the jobs required in the manufacture of the turbines and blades used to build, repair, and replace various installations. In the solar PV sector, indirect employment would include jobs involved in solar panel manufacture and assembly. In both cases, employment might be new jobs or, if a factory is retooled to supply parts to the green energy industry, retained jobs. Under the initial GEA, a domestic content, or made-in-Ontario, rule stipulated that 50 percent of wind power components and 60 percent of solar PV components must be made or assembled in Ontario (OPA 2012). The thinking behind this was that a large number of manufacturing jobs would be created in Ontario to meet the requirement for wind and solar power components; it was seemingly validated by claims of up to 31,000 new jobs created across the province, although no real analysis was ever provided of what form those jobs took (Stinson 2013). After the WTO ruled against these requirements, the domestic content requirement for wind power was reduced to 20 percent, while the requirement for solar PV now ranges between 19 percent (for concentrated PV) and 28 percent (for thin-film solar technology) (OPA 2013). It is still uncertain what influence this ruling will have on future renewable energy projects, although demand for the program seems strong, with almost 2,000 applicants representing 500 MW competing in the most recent round of contracts (OPA 2014a).

An even bigger question looms about the future of existing manufacturing facilities that were intended to service these projects (Blackwell 2013) as the reduction in domestic content rules will make it more difficult for these companies to compete. At the current time, it is impossible to accurately estimate the new or retained indirect employment benefit that might accrue from the manufacture or assembly of renewable energy components.

In addition, farmers who choose to host wind turbines receive rent or royalties on the installations. The skilled trades that service the installations would benefit proportionally to the time spent. In both cases, this might be interpreted as retained indirect employment associated with the sector. No data exist to track this contribution to employment, however, and thus it cannot be included here.

The simple multipliers determined through this review were applied to all wind and solar PV projects being discussed, built, or operated in the Kingston region. The results of this exercise are shown in Table 10.3 below. There is significant potential for employment on a temporary basis during the construction phase. The three operating renewable energy projects around Kingston have created as many as 815 person-years of work. Another 570 person-years of work would be created if all currently approved projects were built as planned, and proposed projects could add a massive number – 1,250 person-years of work – in the region. These figures are particularly significant when one considers the size of the Kingston workforce. Currently, Kingston has a population of 135,900, with an employed labour force of 81,900 and an unemployment rate of 6.2 percent (Statistics Canada 2014b). Temporary construction work could bolster existing jobs or create new opportunities for unemployed workers in the city.

TABLE 10.3
Estimated New Direct and Temporary Employment Associated with Solar and Wind Farm Projects in the Kingston Region as of 2013

Name	*Energy type*	*Nameplate (MW)*	*Temporary jobs*	*New direct jobs*
Completed projects				
First Light I	Solar PV	9.1	100	4
First Light II	Solar PV	10.5	115	2
Wolfe Island Wind Farm	Wind	198	600	10
			Estimated temporary jobs	*Estimated new direct jobs*
Approved projects				
Little Creek Solar Project	Solar PV	10	110	2
Napanee TS Taylor Kidd Solar Project	Solar PV	10	110	2
Kingston Gardiner TS Odessa Solar Project	Solar PV	10	110	2
Kingston Gardiner Highway 2 South Solar Project	Solar PV	10	110	2
Kingston Gardiner TS Unity Road Solar Project	Solar PV	10	110	2
Ernestown Wind Park	Wind	10	20	2
Proposed projects				
Samsung Sol-luce	Solar PV	100	1,100	30
Amherst Island Wind Farm Project	Wind	75	150	4
Proposed projects (on hold)				
Trillium Power Wind I	Wind*	420	848	25
Windstream Wolfe Island Shoals	Wind*	300	606	18
Totals	**Completed**	**217.6**	**815**	**16**
	Approved	**60**	**570**	**12**
	Proposed	**175**	**1,250**	**34**
	On hold	**720**	**1,454**	**43**

*These proposed installations are offshore wind projects and currently under a moratorium.
Sources: For projects and planned outputs: OPA (2014c); for estimated numbers of jobs: authors' compilation.

New direct jobs associated with wind and solar PV installations are very limited compared to the boom in temporary employment. It is estimated that only 16 permanent jobs are associated with the three completed projects in the Kingston region. Moreover, the quality of these jobs is currently unknown, and thus no comment can be made on their sustainability. The approved projects currently expected to be built in the region may add 12 new positions, while the proposed projects could raise that figure by another 34 jobs. While 62 new jobs are not insignificant in a labour market the size of Kingston, this level of employment is not on par with major employers in the region.

As stated above, we know relatively little about the quality of new direct jobs associated with the renewable energy sector in Kingston, but these positions may attract interested applicants. In comparison with jobs in other energy sectors, such as oil and natural gas, employment in the renewable energy sector can be considered sustainable over the long term and may therefore be favoured over traditional sectors.

Additional uncertainty remains about the new and retained indirect employment associated with the renewable energy sector. It is well known that this contribution can be very significant. For example, GreenField Specialty Alcohols Inc., the largest biofuel company in Canada, operates three plants in Ontario, and they can produce about 10 litres of ethanol per bushel of corn used as feedstock (GreenField Specialty Alcohols 2014). In 2012, Ontario produced about 9.6 tonnes of corn per ha (Ontario Ministry of Agriculture, Food and Rural Affairs 2013); the average farm size in Ontario is about 99 ha, with 1.44 farmers employed (Statistics Canada 2014a). Given these statistics, one can estimate that for a single plant with a capacity of 200 million litres per year – such as the GreenField facility near Prescott, just an hour east of Kingston – the equivalent of 2,000 farmers are engaged each year in producing feedstock for the operation. This level of indirect retained employment can be extremely significant, and better ways of assessing these categories of green jobs must be developed.

The MicroFIT program is also a driver of indirect new and retained jobs. An unpublished project carried out at Queen's University in the summer of 2011 found that the program was driving significant investment in Kingston-area businesses that install and service rooftop solar PV. Of the 96 companies investigated, 40 percent reported being directly involved in solar energy, while 76 percent were engaged in some form of energy generation. It was interesting to note that 77 of the companies interviewed were small, locally owned businesses and that half of the companies that responded had fewer than 10 full-time employees. One-third of the businesses interviewed employed graduates from local college training programs (Napier 2011). The MicroFIT program cannot be discounted as a driver of new green employment.

CONCLUSION

This chapter attempts to determine the impact that Ontario's GEA has had on the creation of green sustainable jobs in the city of Kingston and the surrounding region. A number of important conclusions can be drawn from this exercise.

First, the definition of green jobs is currently nebulous and lacks consistency from jurisdiction to jurisdiction and even from agency to agency. In Ontario – despite the fact that the government stated that green jobs are a desired output of the GEA – no one definition is accepted or adopted in an official fashion. The review of existing issues suggests that categories of green jobs must be established to address key issues of permanence, originality, and quality. A framework of five categories of green jobs is adapted from the authors' previous work and proposed here; these categories include new direct and indirect employment, retained direct and indirect jobs, and temporary work. It is also pointed out that no working definition of green jobs currently includes a quantitative metric (or metrics) to define the degree to which such jobs are sustainable. Incorporating such metrics requires widespread agreement on important goals (such as reductions in GHG emissions) and acceptance of a common definition. The lack of a common definition means that the term *green job* can take on a variety of meanings, greatly reducing its utility as a goal or metric in government policy.

The Kingston region has both significant potential for renewable energy production, based on the natural resources of the area, and a considerable number of renewable energy projects already being implemented, primarily using wind and solar PV technologies. The selection of wind and solar options is largely driven by the existing FIT program, which heavily favours these technologies. Already, 217.6 MW of wind and solar power are installed in the region, with another 60 MW approved and 175 MW proposed. These installations would provide a significant proportion of Kingston's energy requirements, helping the city to meet its goal of increasing its use of renewable energy (in a theoretical fashion, as power is currently not retained locally but rather made available throughout the regional grid).

Temporary construction jobs associated with the renewable energy sector are highly significant, with estimates of the total number of job-years in excess of 2,600 across all projects (with the exception of the offshore wind projects currently on hold). Permanent new direct employment is much more limited, however, with only 62 new jobs added to the local economy. Estimates of the green jobs associated with the renewable energy developments around Kingston are confounded by a number of factors, including changes to the domestic content rules, which will certainly impact the number of manufacturing jobs across the province, and a lack of data describing the flow of goods and services through the local economy. It can be assumed that the indirect employment associated with the renewable energy sector is significant; in the forestry sector, for example, indirect employment is much higher than the number of direct jobs. It can also be assumed that many of the indirect jobs developed in association with the renewable energy sector will be retained jobs, not new employment – for example, farmers who provide feedstock or who rent land for wind turbine installations would benefit in an indirect fashion from renewable electricity generation.

Future work in this area will address the uncertainty around indirect employment. Work will also be carried out on the simple multiplier method in order to improve the accuracy of green job estimates. The quality of direct and indirect jobs, the role of unions in green employment, and the potential for new renewable energy development in the region will also be considered. Finally, an examination of the contrast between green jobs associated with the renewable energy sector and conventional energy jobs will be carried out, using nearby gas-fired power plants as comparators.

REFERENCES

Aguilar, F., R. Hartkamp, W.E. Mabee, and K.E. Skog. 2012. "Wood Energy Markets, 2011–2012." In *Forest Products Annual Market Review 2011–2012*, ed. M. Fonseca and J. Posio, 95–106. New York: United Nations.

Alberta. 2009. *Launching Alberta's Energy Future: Provincial Energy Strategy*. Edmonton: Government of Alberta.

BC (British Columbia) MEMPR (Ministry of Energy, Mines and Petroleum Resources). 2009. *The BC Energy Plan: A Vision for Clean Energy Leadership*. Victoria: Ministry of Energy, Mines and Petroleum Resources.

Blackwell, R. 2013. "Ottawa Appeals WTO Ruling on Ontario's Green-Energy Plan." *Globe and Mail*, 6 February, n.p.

Calvert, K., P. Luciani, and W.E. Mabee. 2014. "Thematic Land-Cover Map Assimilation and Synthesis: The Case of Locating Potential Bioenergy Feedstock in Eastern Ontario, Canada." *International Journal of Geographical Information Science* 28 (2):274–95. doi: 10.1080/13658816.2013.833619.

Cryderman, K. 2014. "Washington to Discuss Health Effects of Keystone XL." *Globe and Mail*, 25 February, n.p.

D. Parsons & Associates. 2009. *Greening the Economy: Transitioning to New Careers*. Toronto: Peel-Halton Workforce Development Group / Toronto Workforce Innovation Group / Workforce Planning Board for York Region and Bradford West Gwillimbury.

Earley, S., and W.E. Mabee. 2011. "The Impact of Bioenergy and Biofuel Policies on Employment in Canada." Presented at the Work in a Warming World (W3) Researchers' Workshop: "Greening Work in a Chilly Climate," York University, Toronto, November.

ECO Canada (Environmental Careers Organization). 2010. *Defining the Green Economy*. Labour Market Research Study. Calgary: ECO Canada.

Environment Canada. 2008. "Canadian Wind Energy Atlas." Last updated 29 July. http://www.windatlas.ca/en/nav.php?no=17&field=E1&height=50ANU.

—. 2011. *National Inventory Report 1990–2009: Greenhouse Gas Sources and Sinks in Canada – Executive Summary*. Cat. No.: En81-4/1-2009E-PDF. Ottawa: Environment Canada.

—. 2014. "Progress toward Canada's Greenhouse Gas Emissions Reduction Target." Last modified 11 June. http://www.ec.gc.ca/indicateurs-indicators/default.asp?lang=en&n=CCED3397-1.

Evergreen and ECO Canada. 2012. *The Green Jobs Map: Supplementary Ontario Report*. Labour Market Research Study. Toronto and Calgary: Evergreen / ECO Canada. http://www.evergreen.ca/downloads/pdfs/Green-Jobs-Map-Ontario-2012.pdf.

Focus Kingston. 2010. *Sustainable Kingston: Designing Our Community's Future ... Together*. Sustainable Kingston Plan. Kingston, ON: City of Kingston.

Globe Foundation. 2010. *Securing the Workforce of Tomorrow*. Vancouver: Globe Foundation.

GreenField Specialty Alcohols. 2014. "Locations." http://www.gfsa.com/locations.

Gülen, G. 2011. "Defining, Measuring and Predicting Green Jobs." Lowell, MA: Copenhagen Consensus Center.

IEA (International Energy Agency). 2014. "Statistics: Canada – Balances for 2011." http://www.iea.org/statistics/statisticssearch/report/?country=CANADA&product=balances&year=2011.

IILS (International Institute for Labour Studies). 2011. "Defining 'Green': Issues and Considerations." EC-IILS Joint Discussion Paper Series No. 10. Geneva: International Labour Organization / IILS.

IPCC (Intergovernmental Panel on Climate Change). 2013. "Summary for Policymakers." In *Climate Change 2013: The Physical Science Basis – Contribution of Working Group I to the Fifth Assessment Report of the Intergovernmental Panel on Climate Change,* ed. T.F. Stocker, D. Qin, G.K. Plattner, M. Tignor, S.K. Allen, J. Boschung, A. Nauels, Y. Xia, V. Bex, and P.M. Midgley, 1–28. Cambridge: Cambridge University Press.

Katz, J. With contributions by M. Saltmiras, R. Stanga, R. Sly, S. Kachmar, R. Sheffield, S. Higuchi, and E. Shi. 2012. *Emerging Green Jobs in Canada: Insights for Employment Counsellors into the Changing Labour Market and Its Potential for Entry-Level Employment*. Toronto: Green Skills Network.

Lapointe, M. 2013. "Oilsands Energy Products May Not Be Welcome in European Union." *Hill Times* (Ottawa), 29 July, n.p.

Levin, R., S. Krigstin, and S. Wetzel. 2011. "Biomass Availability in Eastern Ontario for Bioenergy and Wood Pellet Initiatives." *Forestry Chronicle* 87 (1):33–41.

Mabee, W.E., J. Mannion, and T. Carpenter. 2012. "Comparing the Feed-in Tariff Incentives for Renewable Electricity in Ontario and Germany." *Energy Policy* 40:480–89. doi: 10.1016/j.enpol.2011.10.052.

Mabee, W.E., and J. Mirck. 2011. "A Regional Evaluation of Potential Bioenergy Production Pathways in Eastern Ontario, Canada." *Annals of the Association of American Geographers* 101 (4):897–906. doi: 10.1080/00045608.2011.568878.

Manitoba Education. 2012. *Making a Living, Sustainably: Green Jobs and Sustainability Careers*. Winnipeg: Manitoba Education.

Moore, S., V. Durant, and W.E. Mabee. 2013. "Determining Appropriate Feed-in Tariff Rates to Promote Biomass-to-Electricity Generation in Eastern Ontario, Canada." *Energy Policy* 63:607–13. doi: 10.1016/j.enpol.2013.08.076.

Mousseau, N. 2013. With G. Versailles, M.-S. Villeneuve, Y. Dutil, and L. Bessner. *From Greenhouse Gas Reduction to Québec's Energy Self-Sufficiency*. Consultation Paper. Quebec City: Commission sur les enjeux énergétiques du Québec.

Napier, L. 2011. "Employment and Work in the Renewable Energy Industry in the Kingston Region." Unpublished BA (Honours) research report, Department of Geography, Queen's University, Kingston, ON.

Natural Resources Canada. 2014a. "Photovoltaic Potential and Solar Resource Maps of Canada." Last modified 27 November. http://pv.nrcan.gc.ca/index.php?n=1432&m=u&lang=e.

—. 2014b. "State of Canada's Forests: Annual Report 2014." Ottawa: Natural Resources Canada.

Nguyen, H.T., and J.M. Pearce. 2010. "Estimating Potential Photovoltaic Yield with *r.sun* and the Open Source Geographical Resources Analysis Support System." *Solar Energy* 84 (5):831–43. doi: 10.1016/j.solener.2010.02.009.

OECD (Organisation for Economic Co-operation and Development). 2012. *OECD Employment Outlook 2012*. Paris: OECD Publishing.

O'Meara, J. 2013. "Wolfe Island Provides Look into the Life of Wind Farms." *Northumberland News*, 18 January, n.p.

Ontario. 2009. *Bill 150, Green Energy and Green Economy Act, 2009*. Toronto: Government of Ontario.

Ontario. Ministry of Agriculture, Food and Rural Affairs. 2013. "Grain Corn: Area and Production, by County, 2012." Last modified 13 November. http://www.omafra.gov.on.ca/english/stats/crops/ctygrcorn12.htm.

Ontario. Ministry of Energy. 2013. *Achieving Balance: Ontario's Long-Term Energy Plan*. Toronto: Ministry of Energy.

Ontario. Ministry of Energy and Infrastructure. 2011. *Results-Based Plan Briefing Book, 2010–11*. Toronto: Ministry of Energy and Infrastructure.

OPA (Ontario Power Authority). 2012. "Domestic Content." http://fit.powerauthority.on.ca/program-resources/faqs/domestic-content.

—. 2013. "Changes to Domestic Content and New FIT/MicroFIT Price Schedule." 16 August. http://fit.powerauthority.on.ca/newsroom/august-16-2013-program-update.

—. 2014a. "FIT 3.0 Application Summary." Toronto: OPA.

—. 2014b. "FIT Program: Claiming Priority Points." http://fit.powerauthority.on.ca/fit-program/claiming-priority-points.

—. 2014c. "FIT Program: Contract Offers." http://fit.powerauthority.on.ca/program-updates/contract-offers.

Sims, R.E.H., W.E. Mabee, J.N. Saddler, and M. Taylor. 2010. "An Overview of Second Generation Biofuel Technologies." *Bioresource Technology* 101 (6):1570–80. doi 10.1016/j.biortech.2009.11.046.

SkyPower Global. 2014. "First Light II." http://www.skypower.com/skypower-first-light2.php.

Sommers, D. 2013. "BLS Green Jobs Overview." *Monthly Labor Review* 136 (1):3–16.

Statistics Canada. 2014a. "2011 Census of Agriculture." Last modified 28 November. http://www.statcan.gc.ca/ca-ra2011/index-eng.htm.

—. 2014b. "Labour Force Characteristics, Seasonally Adjusted, by Census Metropolitan Area (3 Month Moving Average) (Kingston (Ont.), Peterborough (Ont.), Oshawa (Ont.))." Last modified 7 November. http://www.statcan.gc.ca/tables-tableaux/sum-som/l01/cst01/lfss03e-eng.htm.

Stinson, S. 2013. "Kathleen Wynne Backing Away from McGuinty's Ontario Green Energy Act." *National Post*, 24 June, n.p.

St. Lawrence College. n.d. "Program List." http://www.stlawrencecollege.ca/programs-and-courses/full-time/program-list/.

SunEdison. 2011. "Final Public Meeting for the Napanee TS Taylor Kidd Solar Photovoltaic Project." Toronto: SunEdison.

—. 2014. "First Light 1: First Light Ontario – Canada's First Fully Active Solar PV Energy Park." http://www.sunedison.ca/first-light--ontario-canada.php.

SWITCH. n.d. "About SWITCH." http://www.switchontario.ca/about-SWITCH.

TransAlta. 2014. "Wolfe Island." Last updated 23 May. http://www.transalta.com/facilities/plants-operation/wolfe-island.

UNFCCC (United Nations Framework Convention on Climate Change). 2009. Copenhagen Climate Change Conference, United Nations Framework Convention on Climate Change, December.

US (United States). Department of Commerce. 2010a. *Measuring the Green Economy*. Washington, DC: Department of Commerce, Economics and Statistics Administration.

—. 2010b. "U.S. Carbon Dioxide Emissions and Intensities over Time: A Detailed Accounting for Industries, Government and Households – Technical Appendix." Washington, DC: Department of Commerce, Economics and Statistics Administration.

Wiginton, L.K., H.T. Nguyen, and J.M. Pearce. 2010. "Quantifying Rooftop Solar Photovoltaic Potential for Regional Renewable Energy Policy." *Computers, Environment and Urban Systems* 34 (4):345–57. doi: 10.1016/j.compenvurbsys.2010.01.001.

Worldwatch Institute. 2008. *Green Jobs: Towards Decent Work in a Sustainable, Low-Carbon World*. Nairobi: United Nations Environment Programme.

ABOUT THE CONTRIBUTORS

Geoffrey Bickerton is research director of the Canadian Union of Postal Workers and associate director of the Work in a Warming World research program.

Kean Birch is an assistant professor in the Business and Society program at York University in Toronto. He is the author of a new book, *We Have Never Been Neoliberal: A Manifesto for a Doomed Youth*, due to be published by Zero Books in February 2015.

John Calvert is an associate professor in the Faculty of Health Sciences at Simon Fraser University in Burnaby, British Columbia. His research addresses the impact of climate change on the construction industry, with a particular focus on low-carbon skills and training.

Marjorie Griffin Cohen is a professor in the Department of Political Science and Gender, Sexuality, and Women's Studies at Simon Fraser University. She has written extensively in the areas of political economy and public policy.

John Holmes is a professor emeritus in the Department of Geography at Queen's University in Kingston, Ontario. His primary research interests focus on the restructuring and reorganization of production and work in North America, especially in the automotive industry.

Carla Lipsig-Mummé is a professor in the Work and Labour Studies program at York University. Her recent research focuses on the interaction between the environment and work. She is also editor of the Routledge Studies in Climate, Work and Society series.

Warren Edward Mabee is an associate professor in the Department of Geography at Queen's University and holds the Canada Research Chair in Renewable Energy Development and Implementation. His research focuses on the intersection between renewable energy technology and policy.

Megan MacCallum is an MA candidate in the Department of Geography at Queen's University. She is working on a thesis entitled "The Economic Effects of Renewable and Sustainable Energy Production in the Kingston Region," which draws on the intersection among environmental policy, green energy, and green jobs in eastern Ontario.

Stephen McBride is a professor of political science and Canada Research Chair in Public Policy and Globalization at McMaster University in Hamilton, Ontario. His current research focuses on the political economy of austerity.

Simon Milne is a professor of tourism at Auckland University of Technology and director of the New Zealand Tourism Research Institute. Much of his recent research has focused on how the tourism industry is adapting to environmental change.

Lindsay Napier holds a BA (Honours) degree in human geography and English literature from Queen's University. During her studies, she pursued research on the economic impacts of the emerging renewable energy sector on a localized economy. After graduating in 2012, she began working for the mining industry in the area of corporate social responsibility.

John Shields is a professor in the Department of Politics and Public Administration at Ryerson University in Toronto. His recent research has focused on various aspects of work and labour market development, including green jobs. He is past managing editor of the *Journal of International Migration and Integration*, published by Springer.

Scott Sinclair is a senior research fellow with the Canadian Centre for Policy Alternatives (CCPA), where he directs the centre's Trade and Investment Research Project. Before joining the CCPA, he was a senior trade policy advisor with the government of British Columbia. He has written widely on the impacts of trade treaties on public services and public interest regulation, including *Saving the Green Economy: Ontario's Green Energy Act and the WTO*, published by the CCPA in 2013.

Stephanie Tombari is a PhD candidate in the Department of Political Science at McMaster University. Her current research examines the "green publicity state" in naturalizing neoliberal policies since the global financial crisis. Her next project will examine environmental deregulation in Canada and the United States.

Stuart Trew is senior editor of the *CCPA Monitor*, a monthly journal of the Canadian Centre for Policy Alternatives. Before that, he was a researcher and trade campaigner with the Council of Canadians. During his time at the council, he published reports and academic papers on the free trade regime, social and economic inequality, and climate change.

Steven Tufts is an associate professor in and chair of the Department of Geography at York University. A labour geographer, his current research is on organized labour's response to populism.

Mark Winfield is an associate professor in the Faculty of Environmental Studies at York University. He is also co-chair of the Faculty of Environmental Studies Sustainable Energy Initiative and has published extensively on sustainable energy issues.

Dalton Wudrich, P. Eng. is an MA candidate in environmental studies at York University, studying urban planning in local political economies. He also has interests in municipal finance, infrastructure, and housing.

INDEX

Queen's Policy Studies
Recent Publications

The Queen's Policy Studies Series is dedicated to the exploration of major public policy issues that confront governments and society in Canada and other nations.

Manuscript submission. We are pleased to consider new book proposals and manuscripts. Preliminary inquiries are welcome. A subvention is normally required for the publication of an academic book. Please direct questions or proposals to the Publications Unit by email at spspress@queensu.ca, or visit our website at: www.queensu.ca/sps/books, or contact us by phone at (613) 533-2192.

Our books are available from good bookstores everywhere, including the Queen's University bookstore (http://www.campusbookstore.com/). McGill-Queen's University Press is the exclusive world representative and distributor of books in the series. A full catalogue and ordering information may be found on their web site (**http://mqup.mcgill.ca/**).

For more information about new and backlist titles from Queen's Policy Studies, visit http://www.queensu.ca/sps/books.

School of Policy Studies

Lord Beaconsfield and Sir John A. Macdonald: A Political and Personal Parallel, Michel W. Pharand (ed.) 2015. ISBN 978-1-55339-438-9

Canadian Public-Sector Financial Management, Second Edition, Andrew Graham 2014. ISBN 978-1-55339-426-6

The Multiculturalism Question: Debating Identity in 21st-Century Canada, Jack Jedwab (ed.) 2014. ISBN 978-1-55339-422-8

Government-Nonprofit Relations in Times of Recession, Rachel Laforest (ed.) 2013. ISBN 978-1-55339-327-6

Intellectual Disabilities and *Dual Diagnosis: An Interprofessional Clinical Guide for Healthcare Providers,* Bruce D. McCreary and Jessica Jones (eds.) 2013. ISBN 978-1-55339-331-3

Rethinking Higher Education: Participation, Research, and Differentiation, George Fallis 2013. ISBN 978-1-55339-333-7

Making Policy in Turbulent Times: Challenges and Prospects for Higher Education, Paul Axelrod, Roopa Desai Trilokekar, Theresa Shanahan, and Richard Wellen (eds.) 2013. ISBN 978-1-55339-332-0

Building More Effective Labour-Management Relationships, Richard P. Chaykowski and Robert S. Hickey (eds.) 2013. ISBN 978-1-55339-306-1

Navigationg on the Titanic: Economic Growth, Energy, and the Failure of Governance, Bryne Purchase 2013. ISBN 978-1-55339-330-6

Measuring the Value of a Postsecondary Education, Ken Norrie and Mary Catharine Lennon (eds.) 2013. ISBN 978-1-55339-325-2

Immigration, Integration, and Inclusion in Ontario Cities, Caroline Andrew, John Biles, Meyer Burstein, Victoria M. Esses, and Erin Tolley (eds.) 2012. ISBN 978-1-55339-292-7

Diverse Nations, Diverse Responses: Approaches to Social Cohesion in Immigrant Societies, Paul Spoonley and Erin Tolley (eds.) 2012. ISBN 978-1-55339-309-2

Making EI Work: Research from the Mowat Centre Employment Insurance Task Force, Keith Banting and Jon Medow (eds.) 2012. ISBN 978-1-55339-323-8

Managing Immigration and Diversity in Canada: A Transatlantic Dialogue in the New Age of Migration, Dan Rodríguez-García (ed.) 2012. ISBN 978-1-55339-289-7

International Perspectives: Integration and Inclusion, James Frideres and John Biles (eds.) 2012. ISBN 978-1-55339-317-7

Dynamic Negotiations: Teacher Labour Relations in Canadian Elementary and Secondary Education, Sara Slinn and Arthur Sweetman (eds.) 2012. ISBN 978-1-55339-304-7

Where to from Here? Keeping Medicare Sustainable, Stephen Duckett 2012. ISBN 978-1-55339-318-4

International Migration in Uncertain Times, John Nieuwenhuysen, Howard Duncan, and Stine Neerup (eds.) 2012. ISBN 978-1-55339-308-5

Centre for International and Defence Policy

Afghanistan in the Balance: Counterinsurgency, Comprehensive Approach, and Political Order, Hans-Georg Ehrhart, Sven Bernhard Gareis, and Charles Pentland (eds.), 2012. ISBN 978-1-55339-353-5

Institute of Intergovernmental Relations

Canada: The State of the Federation 2011, Nadia Verrelli (ed.), 2014. ISBN 978-1-55339-207-1

Canada and the Crown: Essays on Constitutional Monarchy, D. Michael Jackson and Philippe Lagassé (eds.), 2013. ISBN 978-1-55339-204-0

Paradigm Freeze: Why It Is So Hard to Reform Health-Care Policy in Canada, Harvey Lazar, John N. Lavis, Pierre-Gerlier Forest, and John Church (eds.), 2013. ISBN 978-1-55339-324-5

Canada: The State of the Federation 2010, Matthew Mendelsohn, Joshua Hjartarson, and James Pearce (eds.), 2013. ISBN 978-1-55339-200-2

The Democratic Dilemma: Reforming Canada's Supreme Court, Nadia Verrelli (ed.), 2013. ISBN 978-1-55339-203-3